progress in filtration and separation

3

progress in filtration and separation

3

edited by

r.j. wakeman

Department of Chemical Engineering,
University of Exeter, Exeter (Great Britain)

ELSEVIER SCIENTIFIC PUBLISHING COMPANY

Amsterdam — Oxford — New York 1983

ELSEVIER SCIENTIFIC PUBLISHING COMPANY
Molenwerf 1
P.O. Box 211, 1000 AE Amsterdam, The Netherlands

Distributors for the United States and Canada:

ELSEVIER SCIENCE PUBLISHING COMPANY, INC.
52, Vanderbilt Avenue
New York, N.Y. 10017

ISBN 0-444-42168-8 (Vol. 3)
ISBN 0-444-41820-2 (Series)

Printed in The Netherlands

INTRODUCTION

The aims of this multi-author book remain the same as with previous volumes; to provide researchers and engineers with critical reviews and carefully considered opinions of recent works in the field of filtration and separation and adjoining areas. Whilst some chapters are intended to bring older topics up to date, others gather and examine the results of new or freshly utilized techniques and methods of seeming promise. As before, the authors have provided definitive reports giving their own analyses of the subjects, not merely annotated bibliographies. Individual authors have exercised their judgement as to which are significant papers in their own areas of expertise, and all have contributed significantly to the research about which they write.

A different approach to the fundamental theory of filtration is adopted by Professor Willis who describes the development of the material and momentum balance equations for multiphase mixtures, and shows how these equations can be simplified and applied to filtration. The equations are capable of describing a Newtonian fluid phase which contains a particulate phase which may be elastic, non-deformable, or soluble in the liquid.

Drs. Bloor and Ingham bring together the results of extensive theoretical analyses of various aspects of the fluid mechanics of hydrocyclones in their chapter. Having developed a good picture of the fluid mechanics of the system which enables prediction of the fluid velocity distribution, a method for calculating the separating efficiency for small dense particles is given and some insight into the mechanisms of blockage at high solids concentrations is provided.

The general principle, operating characteristics, and performance data of dielectrophoretic filtration and separation processes are described in the article by Professors Lin and Benguigui. High gradient electronic separation and dielectric filtration are based on the use of a polarization force exerted by a non-uniform electric field (dielectrophoretic effects). The requirements of such filter systems and their possible areas of application are considered in some depth; this contribution is complementary to those by Dr. Andres (p. 125, volume 2) and by Professor Birss and Dr. Parker (p. 171, volume 2).

Dr. Perkin demonstrates the principles of drying porous materials with electromagnetic energy generated at radio and microwave frequencies, dielectric heating. Current developments and industrial applications of the technique are discussed, along with methods of power generation and its application and the drying characteristics derived from theoretical modelling.

Richard J. Wakeman

Department of Chemical Engineering,
University of Exeter,
Exeter, Devon, U.K.

CONTENTS

A MULTIPHASE THEORY OF FILTRATION

M. S. Willis
Chemical Eng. Dept., Univ. of Akron, Akron, OH (United States)

CONTENTS

ABSTRACT

The volume averaging procedure, operators, deviations, and averaging theorems are developed and then used to derive the multiphase equations that provide the basis for the filtration theory. The permeability function is introduced into the multiphase equations together with constitutive equations for both phases. These equations describe a Newtonian fluid phase which contains particulate phases that are either Hookean elastic, or non-deformable, or non-deformable but soluble in the liquid phase.

Dimensional analysis of the equations for an insoluble, non-deformable particulate phase in a Newtonian liquid phase indicates that the dominant forces are the pressure, gravity and interfacial drag. This information is used to simplify the multiphase equations.

The simplified results of the dimensional analysis for a non-deformable particulate phase are applied to a one-dimensional axial filtration and the analysis, which includes experimental verification and a parametic study, shows that the permability at the cake septum interface influences the filtrate rate, pumping cost, design, and operation of cake filters but is independent of the slurry concentration, pressure, and initial deposition rate.

INTRODUCTION

Filtration is a special case of a much larger class of multipase phenomena. Due to the complexities of these multiphase systems, attempts to describe them are usually intuitive and empirical. This is also true in filtration where the empirically based series resistance model (ref. 1,2) of cake filtration predicts a linear relation between reciprocal rate and filtrate volume with the slope related to the filter cake resistivity and the intercept proportional to the septum resistance.

The major disadvantages of empirical models are that they are usually limited in application to specific multiphase systems, have a narrow range of validity, must be independently verified for each substance and, hence, require a considerable amount of experimental effort.

On the other hand, any development based on fundamental principles applies to the set of all substances that meet the assumptions used to develop the theory. For example, the multiphase theory of filtration (ref. 3) is based on fundamental principles and is limited to one dimensional filtrations of slurries for which the liquid phase is Newtonian and the particulate phase is non-deformable. There are, of course, a considerable number of liquid and particulate combinations that meet these restrictions. If the theory is verified for one member of the set, then it is verified for every member of the set. Consequently, the effort used to develop the theory is compensated for by a decrease in experimental work.

In this chapter, the averaging procedure, theorems and development of the multiphase equations are described. Constitutive equations are incorporated into the general equations and then these multiphase equations are applied to slurries composed of Newtonian liquids and particulate phases that are soluble or elastic or non-deformable. The latter sections describe the experimental verification of the one-dimensional multiphase filtration theory for slurries of Newtonian fluids with non-deformable particulate phases.

FORMULATION OF THE MULTIPHASE EQUATIONS

Averaging Procedure

The averaging procedure to be described can be applied to multiphase systems in which all the phases are isolated from each other (ref. 4,5). However, in filtration a particulate phase is dispersed in a continuous fluid phase and this is the case for which the multiphase equations will be developed here.

The particles of the dispersed solid particulate phase have a characteristic length ℓ and the volume over which averages are taken has a characteristic length D. The filter cake has a characteristic length L. In order to obtain meaningful averages, the averaging volume must be much larger than ℓ and much smaller than L. If the averaging volume is about the same size as ℓ, then the averaged quantities will fluctuate and if the averaging volume is about the size L, then spatial variations cannot be taken into account. For most filtrations these characteristic lengths can be identified and the averaging procedure is valid.

The averaging volume is denoted by V_D since the characteristic length is D and the surface area is A_D. The orientation, shape, and size of the averaging volume is the same at all locations and for all times. The volumes of the fluid and particulate phases within the averaging volume, however, can change with location and time. The interfacial area between the phases is denoted by $A_{\alpha\beta}$. The properties that are averaged over V_D are independent of the size of the averaging volume (or area) and are continuous functions of space and time.

The subscript α indicates that the property or averages are associated with the continuous fluid phase and β indicates that the property or averages are associated with the particulate phase.

The centroid of the averaging volume is located at a position $\underline{x}$ relative to an inertial frame of reference as shown in Fig. 1. Any point in the averaging volume, V_D, is located with respect to the inertial frame by the position vector $\underline{r}$ and with respect to the centroid of the averaging volume by $\underline{\xi}$ such that

$$\underline{r} = \underline{x} + \underline{\xi} \tag{1}$$

The distribution of the two phases within the averaging volume is described by the distribution function

$$\gamma_\alpha = \gamma_\alpha(\underline{r},t) = \begin{cases} 1, \text{ if } \underline{r} \in V_\alpha \\ 0, \text{ if } \underline{r} \in V_\beta \end{cases} \qquad (2)$$

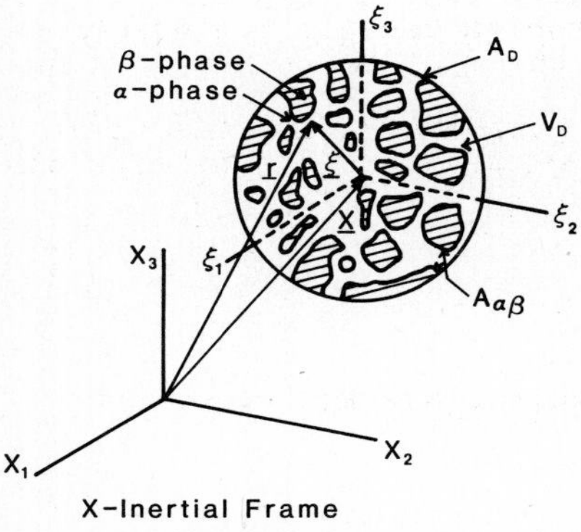

Fig. 1. Inertial and centroid reference frames for the averaging volume, V_D, with surface area, A_D, containing a continuous α-phase and a dispersed β-phase.

On the interfaces between the two phases γ_α is not defined but its right and left hand limits exist there. The function not only describes the distribution of the phases but it also counts the points over which the average is taken.

The average value is obtained by integrating over all points of the averaging volume by allowing $\underline{\xi}$ to vary and assigning the value of the integral to the centroid of the averaging volume. The value of the integral is a function of $\underline{x}$ and time but it is independent of the limits of the averaging volume.

The volume of the **α**-phase in the averaging volume is given by

$$V_\alpha(\underline{x},t) = \int_{V_D} \gamma_\alpha(\underline{x}+\underline{\xi},t)\,d_\xi V \qquad (3)$$

where $d_\xi V$ indicates that the integration is over the averaging volume in the ξ-coordinate system. The portion of the surface of the averaging volume occupied by the **α**-phase is

$$A_\alpha(\underline{x},t) = \int_{A_D} \gamma_\alpha(\underline{x}+\underline{\xi},t)\,d_\xi A \qquad (4)$$

and the volume fraction of the **α**-phase is given by

$$\varepsilon_\alpha(\underline{x},t) = \frac{1}{V_D}\int_{V_D} \gamma_\alpha(\underline{r},t)\,d_\xi V \qquad (5)$$

where

$$\varepsilon_\alpha + \varepsilon_\beta = 1 \tag{6}$$

Averaging Operators

Interstitial quantities are difficult or impossible to measure but averaging circumvents this problem because the quantities that result from the averaging operations must be quantities that can be measured.

Let f represent an arbitrary property of any tensorial order. Then the following three averaging operators can be defined. The first is the volume average operator, $\langle\ \rangle_\alpha$, given by

$$\langle f\rangle_\alpha(\underline{x},t) \equiv \frac{1}{V_D}\int_{V_D} f(\underline{r},t)\gamma_\alpha(\underline{r},t)d_\xi V \tag{7}$$

and the second is the intrinsic volume average operator, $\langle\ \rangle_\alpha^\alpha$, given by

$$\langle f\rangle_\alpha^\alpha(\underline{x},t) \equiv \frac{1}{V_\alpha}\int_{V_D} f(\underline{r},t)\gamma_\alpha(\underline{r},t)d_\xi V \tag{8}$$

These two averages are related by

$$\langle f\rangle_\alpha = \varepsilon_\alpha\langle f\rangle_\alpha^\alpha \tag{9}$$

As an example of these two averages consider the density. The volume average density

$$\langle \rho\rangle_\alpha = \frac{1}{V_D}\int_{V_D}\rho\gamma_\alpha d_\xi V \tag{10}$$

is the bulk density. (Henceforth, the dependence on $\underline{x}$ and t and on $\underline{r}$ and t is understood.) The intrinsic volume average density is

$$\langle \rho\rangle_\alpha^\alpha = \frac{1}{V_\alpha}\int_{V_D}\rho\gamma_\alpha d_\xi V \tag{11}$$

If the density is constant, then

$$\langle \rho\rangle_\alpha^\alpha = \rho \tag{12}$$

and the bulk density is

$$\langle \rho\rangle_\alpha = \varepsilon_\alpha\rho \tag{13}$$

The third averaging operator is the mass average operator, $\bar{\ }^\alpha$, given by

$$\bar{f}^\alpha = \frac{1}{\langle\rho\rangle_\alpha V_D}\int_{V_D}\rho f\gamma_\alpha d_\xi V \tag{14}$$

Hassanizadeh and Gray (ref. 4) discuss these averaging operators in more detail and they also define an area average operator.

The Dirac Delta Function

To obtain the averaged form of a balance equation the averages of spatial and time derivatives must be converted to spatial and time derivatives of averaged quantites. Before these two theorems can be developed, the distribution function must be related to the Dirac delta function (ref. 6).

The distribution function, given in Equation (2), can be written in one dimension as the sum of unit step functions

$$\gamma_\alpha(x) = H(x-a_0) - H(x-a_1) + H(x-a_2) + H(x-a_3) \tag{15}$$

where

$$H(x-a) = \begin{cases} 0, & x < a \\ 1, & x > a \end{cases} \tag{16}$$

is the unit step function (ref. 7) and

$$\frac{d}{dx}H(x-a) = \delta(x-a) \tag{17}$$

where $\delta(x-a)$ is the Dirac delta function (ref. 8).

By taking the derivative of Equation (15), using Equation (17), and generalizing to three dimensions, the following relation

$$\underline{\nabla}_x \gamma_\alpha(\underline{x}) = - \underline{n}_{\alpha\beta}\delta(\underline{x} - \underline{x}_{\alpha\beta}) \tag{18}$$

is obtained for the gradient of the distribution function. The unit normal vector, $\underline{n}_{\alpha\beta}$, is positive when it points outward from the α-phase, $\underline{x}$ is a position vector and $\underline{x}_{\alpha\beta}$ is the position vector of the $\alpha\beta$-interface.

For volume averaging, there are two gradient operators. One is associated with the x-coordinate system and one with the ξ-coordinate system. Since the ξ-coordinate system is simply displaced without rotation from the x-coordinate system, then a point in the two coordinate systems is given by $x = \xi + C$, where C is the displacement of the ξ-coordinate system from the x-coordinate system. Consequently, the gradient operator is the same in both coordinate systems and the subscript on the gradient operator can be dropped.

If the β-phase moves, then γ_α is a function of time as well as position and the total derivative of γ_α with respect to time is

$$\frac{d\gamma_\alpha}{dt} = \frac{\partial\gamma_\alpha}{\partial t} + \sum_{i=1}^{i=3} \frac{\partial\gamma_\alpha}{\partial x_i}\frac{dx_i}{dt} \tag{19}$$

Since $\partial\gamma_\alpha/\partial x_i$ is non-zero only on the $\alpha\beta$-interfaces, then the dx_i/dt are the velocity components of the interface. If the motion of the interface is followed, γ_α is constant there and the total derivative of γ_α is zero. Hence

$$\frac{\partial \gamma_\alpha}{\partial t} = -\underline{w}\cdot\underline{\nabla}\gamma_\alpha \qquad (20)$$

where $\underline{w}$ is the velocity of the $\alpha\beta$-interface.

Equations (18) and (20) allow the spatial and time deriatives of the distribution function to be written in terms of the Dirac delta function. When the Dirac delta funtion appears in the integrand of a volume averaging integral, it selects from all the points in the volume only those that are on the interfaces between the two phases.

Average of the Divergence and of the Gradient

The average of the divergence is obtained by letting $f = \underline{\nabla}\cdot\underline{\psi}$ in Equation (7) and then using the relation

$$(\underline{\nabla}\cdot\underline{\psi})\gamma_\alpha = \underline{\nabla}\cdot(\underline{\psi}\gamma_\alpha) - \underline{\psi}\cdot\underline{\nabla}\gamma_\alpha \qquad (21)$$

to obtain

$$\frac{1}{V_D}\int_{V_D}(\underline{\nabla}\cdot\underline{\psi})\gamma_\alpha d_\xi V = \frac{1}{V_D}\int_{V_D}\underline{\nabla}\cdot(\underline{\psi}\gamma_\alpha)d_\xi V - \frac{1}{V_D}\int_{V_D}\underline{\psi}\cdot\underline{\nabla}\gamma_\alpha d_\xi V \qquad (22)$$

Equation (18) is substituted into the second integral on the right and then the selective property of the Dirac delta function is used to obtain

$$\frac{1}{V_D}\int_{V_D}(\underline{\nabla}\cdot\underline{\psi})\gamma_\alpha d_\xi V = \underline{\nabla}\cdot[\frac{1}{V_D}\int_{V_D}\underline{\psi}\gamma_\alpha d_\xi V] + \frac{1}{V_D}\int_{A_{\alpha\beta}}\underline{\psi}\cdot\underline{n}_{\alpha\beta}d_\xi A \qquad (23)$$

where the integration and differention have been interchanged on the right because V_D is independent of position. The fact that the del operators in the x- and ξ-coordinate systems are the same has been used also. The proof for the gradient is analogous with the result

$$\frac{1}{V_D}\int_{V_D}(\underline{\nabla}\psi)\gamma_\alpha d_\xi V = \underline{\nabla}[\frac{1}{V_D}\int_{V_D}\psi\gamma_\alpha d_\xi V] + \frac{1}{V_D}\int_{A_{\alpha\beta}}\psi\underline{n}_{\alpha\beta}d_\xi A \qquad (24)$$

which, for $\psi = 1$, becomes

$$\underline{\nabla}\varepsilon_\alpha = -\frac{1}{V_D}\int_{A_{\alpha\beta}}\underline{n}_{\alpha\beta}d_\xi A \qquad (25)$$

Average of the Time Derivative

The average of the time derivative is obtained by letting $f = \partial\psi/\partial t$ in Equation (7) and then using

$$(\frac{\partial\psi}{\partial t})\gamma_\alpha = \frac{\partial(\psi\gamma_\alpha)}{\partial t} - \psi\frac{\partial\gamma_\alpha}{\partial t} \qquad (26)$$

to obtain

$$\frac{1}{V_D}\int_{V_D}\left(\frac{\partial\psi}{\partial t}\right)\gamma_\alpha d_\xi V = \frac{1}{V_D}\int_{V_D}\frac{\partial(\psi\gamma_\alpha)}{\partial t}d_\xi V - \frac{1}{V_D}\int_{V_D}\psi\frac{\partial\gamma_\alpha}{\partial t}d_\xi V \tag{27}$$

The time derivative of the distribution function in Equation (27) is converted to the Dirac delta function by combining Equations (18) and (20) with the result

$$\frac{1}{V_D}\int_{V_D}\left(\frac{\partial\psi}{\partial t}\right)\gamma_\alpha d_\xi V = \frac{\partial}{\partial t}\left[\frac{1}{V_D}\int_{V_D}\psi\gamma_\alpha d_\xi V\right] - \frac{1}{V_D}\int_{A_{\alpha\beta}}\psi\underline{w}\cdot\underline{n}_{\alpha\beta}d_\xi A \tag{28}$$

where the integration and differention with respect to time can be interchanged because the averaging volume is independent of time and $\underline{w}$ is the interface velocity. If $\psi = 1$ then Equation (28) becomes

$$\frac{\partial\varepsilon_\alpha}{\partial t} = \frac{1}{V_D}\int_{A_{\alpha\beta}}\underline{w}\cdot\underline{n}_{\alpha\beta}d_\xi A \tag{29}$$

which indicates that if the interface velocity is zero then the porosity is not a function of time.

Deviations and Averages of Deviations

In the development of the multiphase equations it is necessary to convert averages of products to products of averages. This is accomplished by the introduction of deviations.

The deviation of the function f at some point $\underline{r}$ in the averaging volume from its mass average value located at the centroid of the averaging volume is

$$\tilde{f}^\alpha(\underline{x}+\underline{\xi},t) = f(\underline{x}+\underline{\xi},t) - \bar{f}^\alpha(\underline{x},t) \tag{30}$$

These deviations are defined over the same set of spatial points that are used to obtain the mass average value of the functions f. The distribution function automatically keeps track of the points over which averages are taken.

Since the mass average value of the function f is uniform over the averaging volume, then the following identities are valid

$$\overline{\tilde{f}^\alpha}^\alpha = \frac{1}{\langle\rho\rangle_\alpha V_D}\int_{V_D}\rho(f - \bar{f}^\alpha)\gamma_\alpha d_\xi V = 0 \tag{31}$$

$$\overline{\tilde{f}^\alpha\bar{h}^\alpha}^\alpha = \overline{\bar{f}^\alpha\tilde{h}^\alpha}^\alpha = 0 \tag{32}$$

$$\overline{fh}^\alpha = \bar{f}^\alpha\bar{h}^\alpha + \overline{\tilde{f}^\alpha\tilde{h}^\alpha}^\alpha \tag{33}$$

Multiphase Mass Equation and Jump Mass Balance

The equation of continuity (ref. 9)

$$\frac{\partial \rho}{\partial t} + \underline{\nabla} \cdot \rho \underline{v} = 0 \tag{34}$$

can be multiplied by the distribution function and then integrated over the averaging volume to obtain

$$\frac{1}{V_D} \int_{V_D} \left(\frac{\partial \rho}{\partial t}\right) \gamma_\alpha d_\xi V + \frac{1}{V_D} \int_{V_D} (\underline{\nabla} \cdot \rho \underline{v}) \gamma_\alpha d_\xi V = 0 \tag{35}$$

The averaging theorems, Equations (23) and (28), can be used to obtain

$$\frac{\partial}{\partial t}\left[\frac{1}{V_D} \int_{V_D} \rho \gamma_\alpha d_\xi V\right] - \frac{1}{V_D} \int_{A_{\alpha\beta}} \rho (\underline{w} - \underline{v}) \cdot \underline{n}_{\alpha\beta} d_\xi A + \underline{\nabla} \cdot \left[\frac{1}{V_D} \int_{V_D} \rho \underline{v} \gamma_\alpha d_\xi V\right] = 0 \tag{36}$$

which, upon the application of Equations (9), (10) and (14), becomes

$$\frac{\partial}{\partial t}(\varepsilon_\alpha \langle \rho \rangle_\alpha^\alpha) + \underline{\nabla} \cdot (\varepsilon_\alpha \langle \rho \rangle_\alpha^\alpha \underline{\bar{v}}^\alpha) + \frac{1}{V_D} \int_{A_{\alpha\beta}} \rho (\underline{v} - \underline{w}) \cdot \underline{n}_{\alpha\beta} d_\xi A = 0 \tag{37}$$

The integral in Equation (37) accounts for the total exchange of mass between the two phases due to a phase change. The term "phase change" includes interphase transport in addition to a change of state such as condensation. If the $\alpha\beta$-interface is a material surface (i.e., there is no flux of mass through it), then the integrand is identically zero.

The conditions at the interfaces in the averaging volume are accounted for by examining a material volume that contains an interface as shown in Fig. 2. The procedure for obtaining the jump

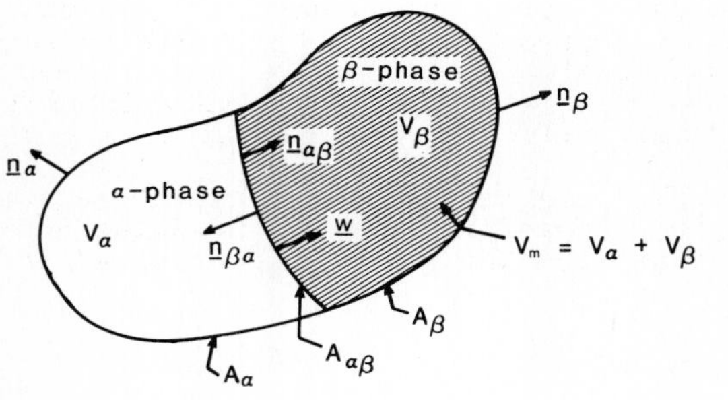

Fig. 2. Material volume, V_m, for the determination of the interfacial jump conditions.

balance at an interface consists of making a balance on each phase, adding these two equations and comparing the result with an overall balance over the material volume, V_m. It is assumed that surface thermodynamic properties, such as surface tension, are absent. The properties of each phase are identified by a subscript.

Integration of the continuity condition, Equation (34), for the α-phase and application of the Reynolds' transport theorem (ref. 10) and the divergence theorem gives

$$\frac{d}{dt}\int_{V_\alpha}\rho_\alpha dV - \int_{A_{\alpha\beta}}\rho_\alpha(\underline{w} - \underline{v}_\alpha)\cdot\underline{n}_{\alpha\beta}dA = 0 \qquad (38)$$

where $\underline{w}$ is the velocity of the interface.

The result for the β-phase is analogous

$$\frac{d}{dt}\int_{V_\beta}\rho_\beta dV - \int_{A_{\beta\alpha}}\rho_\beta(\underline{w} - \underline{v}_\beta)\cdot\underline{n}_{\beta\alpha}dA = 0 \qquad (39)$$

and the sum is

$$\frac{d}{dt}\int_{V_\alpha+V_\beta}(\rho_\alpha + \rho_\beta)dV - \int_{A_{\alpha\beta}}[\rho_\alpha(\underline{w} - \underline{v}_\alpha)\cdot\underline{n}_{\alpha\beta} + \rho_\beta(\underline{w} - \underline{v}_\beta)\cdot\underline{n}_{\beta\alpha}]dA = 0 \qquad (40)$$

The intergal of the continuity condition for both phases over the material volume, V_m, is

$$\int_{V_m}\frac{\partial}{\partial t}(\rho_\alpha + \rho_\beta)dV + \int_{V_m}\underline{\nabla}\cdot(\rho_\alpha + \rho_\beta)\underline{v}dV = 0 \qquad (41)$$

which, after the application of the Reynolds' transport theorem and the divergence theorem, becomes

$$\frac{d}{dt}\int_{V_m}(\rho_\alpha + \rho_\beta)dV = 0 \qquad (42)$$

This equation is then substituted into Equation (40) with the result

$$\rho(\underline{w} - \underline{v})\cdot\underline{n}_{\alpha\beta}|_\alpha + \rho(\underline{w} - \underline{v})\cdot\underline{n}_{\beta\alpha}|_\beta = 0 \qquad (43)$$

since the interfacial area $A_{\alpha\beta}$ is arbitrary. The notation $|_\alpha$ indicates the limit of the preceeding term as the interface is approached from the α-phase side and $\underline{n}_{\alpha\beta} = -\underline{n}_{\beta\alpha}$ as shown in Fig. 2.

Multiphase Species Equation and Jump Species Balance

The fluid α-phase is assumed to contain a solute species, labeled species 2, and a solvent species, labeled species 1. This assumption permits Fick's law to be used and if the fluid α-phase is dilute in the solute species then the density of the fluid phase is approximately equal to the density of the solvent species

$$\rho = \rho_1 + \rho_2 \doteq \rho_1 \qquad (44)$$

and the mass average velocity is

$$\underline{v} = \omega_1\underline{v}_1 + \omega_2\underline{v}_2 \doteq \underline{v}_1 \quad (45)$$

The inventory equation (ref. 9) for each species in the fluid α-phase is

$$\frac{\partial \rho_i}{\partial t} + \underline{\nabla}\cdot\underline{j}_i + \underline{\nabla}\cdot\rho_i\underline{v} = r_i \quad , \; i = 1,2 \quad (46)$$

and the volume average of this equation is

$$\frac{1}{V_D}\int_{V_D}\left(\frac{\partial \rho_i}{\partial t}\right)\gamma_\alpha d_\xi V + \frac{1}{V_D}\int_{V_D}(\underline{\nabla}\cdot\underline{j}_i)\gamma_\alpha d_\xi V + \frac{1}{V_D}\int_{V_D}(\underline{\nabla}\cdot\rho_i\underline{v})\gamma_\alpha d_\xi V =$$

$$\frac{1}{V_D}\int_{V_D} r_i\gamma_\alpha d_\xi V \quad , \; i = 1,2 \quad (47)$$

When the volume averaging theorems, Equations (23) and (28), the deviation definition, Equation (30), and the definition of the mass average, Equation (14), are used in Equation (47), then the result is the α-phase volume averaged multiphase species balance

$$\frac{\partial}{\partial t}\langle\rho\rangle_\alpha\bar{\omega}_i^\alpha + \underline{\nabla}\cdot(\langle\rho\rangle_\alpha\bar{\omega}_i^\alpha\bar{\underline{v}}^\alpha) + \underline{\nabla}\cdot(\langle\underline{j}_i\rangle_\alpha + \langle\rho\rangle_\alpha\overline{\tilde{\omega}_i^\alpha\tilde{\underline{v}}^\alpha}^\alpha) +$$

$$\frac{1}{V_D}\int_{A_{\alpha\beta}}\rho\omega_i(\underline{v} - \underline{w})\cdot\underline{n}_{\alpha\beta}d_\xi A + \frac{1}{V_D}\int_{A_{\alpha\beta}}\underline{j}_i\cdot\underline{n}_{\alpha\beta}d_\xi A = \langle r_i\rangle_\alpha \quad , \; i = 1,2 \quad (48)$$

If the particulate β-phase also has two species present, then an equation analogous to Equation (48) can be written for that phase.

The procedure for obtaining the jump species balance is the same as that for the jump mass balance. The material volume is shown in Fig. 2 and the properties of each phase are again identified with an appropriate subscript. The ith species is assumed to exist in both phases and the sum of the species balances for each phase is

$$\frac{d}{dt}\int_{V_m}(\rho_{i\alpha} + \rho_{i\beta})dV + \int_{A_{\alpha\beta}}[\rho_{i\alpha}(\underline{v}_\alpha - \underline{w})\cdot\underline{n}_{\alpha\beta} + \rho_{i\beta}(\underline{v}_\beta - \underline{w})\cdot\underline{n}_{\beta\alpha}]dA +$$

$$\int_{A_\alpha}\underline{n}_\alpha\cdot\underline{j}_{i\alpha}dA + \int_{A_\beta}\underline{n}_\beta\cdot\underline{j}_{i\beta}dA + \int_{A_{\alpha\beta}}[\underline{n}_{\alpha\beta}\cdot\underline{j}_{i\alpha} + \underline{n}_{\beta\alpha}\cdot\underline{j}_{i\beta}]dA =$$

$$\int_{V_m}(r_{i\alpha} + r_{i\beta})dV \quad , \; i = 1,2 \quad (49)$$

The integral over the material volume V_m is

$$\frac{d}{dt}\int_{V_m}(\rho_{i\alpha} + \rho_{i\beta})dV + \int_{A_\alpha}\underline{n}_\alpha\cdot\underline{j}_{i\alpha}dA + \int_{A_\beta}\underline{n}_\beta\cdot\underline{j}_{i\beta}dA =$$

$$\int_{V_m}(r_{i\alpha} + r_{i\beta})dV \quad , \; i = 1,2 \quad (50)$$

and the combination of Equation (49) and (50) gives the jump species balance

$$[(\rho_i(\underline{v} - \underline{w}) + \underline{j}_i)\cdot\underline{n}_{\alpha\beta}]|_{\alpha} + [(\rho_i(\underline{v} - \underline{w}) + \underline{j}_i)\cdot\underline{n}_{\beta\alpha}]|_{\beta} = 0 \quad , i = 1,2 \tag{51}$$

The notation indicates that the velocity differences in each phase are not equal to each other because of the different diffusive fluxes in each phase.

Multiphase Momentum Equation and Jump Momentum Balance

The linear momentum equation (ref. 9) is

$$\frac{\partial}{\partial t}(\rho\underline{v}) = -\underline{\nabla}\cdot\rho\underline{v}\underline{v} + \underline{\nabla}\cdot\underline{\underline{\pi}} + \rho\underline{g} \tag{52}$$

and the volume average of this equation is

$$\frac{1}{V_D}\int_{V_D}(\frac{\partial}{\partial t}\rho\underline{v})\gamma_\alpha d_\xi V = -\frac{1}{V_D}\int_{V_D}(\underline{\nabla}\cdot\rho\underline{v}\underline{v})\gamma_\alpha d_\xi V + \frac{1}{V_D}\int_{V_D}(\underline{\nabla}\cdot\underline{\underline{\pi}})\gamma_\alpha d_\xi V +$$

$$\frac{1}{V_D}\int_{V_D}\rho\underline{g}\gamma_\alpha d_\xi V \tag{53}$$

The final form of the volume averaged momentum equation

$$\frac{\partial}{\partial t}\langle\rho\rangle_\alpha\bar{\underline{v}}^\alpha + \underline{\nabla}\cdot(\langle\rho\rangle_\alpha\bar{\underline{v}}^\alpha\bar{\underline{v}}^\alpha) = \underline{\nabla}\cdot\langle\underline{\underline{\pi}}\rangle_\alpha + \langle\rho\rangle_\alpha\bar{\underline{g}}^\alpha - \underline{\nabla}\cdot(\langle\rho\rangle_\alpha\overline{\tilde{\underline{v}}^\alpha\tilde{\underline{v}}^\alpha}^\alpha) +$$

$$\frac{1}{V_D}\int_{V_D}\rho\underline{v}(\underline{w} - \underline{v})\cdot\underline{n}_{\alpha\beta}d_\xi A + \frac{1}{V_D}\int_{A_{\alpha\beta}}\underline{\underline{\pi}}\cdot\underline{n}_{\alpha\beta}d_\xi A \tag{54}$$

is obtained by using again the averaging theorems, Equations (23) and (28), the deviation definition, Equation (30), and the definition of the mass average, Equation (14). By replacing the **α** with a **β** in Equation (54), the analogous momentum equation for the **β**-phase is obtained.

The jump momentum balance is developed by using Fig. 2 and the same procedure as that used for the jump mass and jump species balances. The form of the jump momentum balance

$$[(\rho\underline{v}(\underline{w} - \underline{v}) + \underline{\underline{\pi}})\cdot\underline{n}_{\alpha\beta}]|_{\alpha} + [(\rho\underline{v}(\underline{w} - \underline{v}) + \underline{\underline{\pi}})\cdot\underline{n}_{\beta\alpha}]|_{\beta} = 0 \tag{55}$$

is analogous to that of the jump species balance, Equation (51).

It is possible to combine the multiphase continuity condition, Equation (37), with the multiphase momentum equation to obtain

$$\varepsilon_\alpha\langle\rho\rangle_\alpha^\alpha[\frac{\partial}{\partial t}\bar{\underline{v}}^\alpha + \bar{\underline{v}}^\alpha\cdot\underline{\nabla}\bar{\underline{v}}^\alpha] = \underline{\nabla}\cdot\langle\underline{\underline{\pi}}\rangle_\alpha + \varepsilon_\alpha\langle\rho\rangle_\alpha^\alpha\bar{\underline{g}}^\alpha - \underline{\nabla}\cdot(\varepsilon_\alpha\langle\rho\rangle_\alpha^\alpha\overline{\tilde{\underline{v}}^\alpha\tilde{\underline{v}}^\alpha}^\alpha) +$$

$$\frac{1}{V_D}\int_{A_{\alpha\beta}}\rho\tilde{\underline{v}}^\alpha(\underline{w} - \underline{v})\cdot\underline{n}_{\alpha\beta}d_\xi A + \frac{1}{V_D}\int_{A_{\alpha\beta}}\underline{\underline{\pi}}\cdot\underline{n}_{\alpha\beta}d_\xi A \tag{56}$$

CONSTITUTIVE EQUATIONS FOR THE MULTIPHASE EQUATIONS

Newtonian Fluid Phase

The stress tensor in Equation (56) can be written as

$$\langle \underline{\underline{\pi}} \rangle_\alpha = \langle \underline{\underline{\tau}} \rangle_\alpha - \langle P\underline{\underline{I}} \rangle_\alpha \tag{57}$$

where $\underline{\underline{I}}$ is the identity tensor. The surface forces can then be written as

$$\underline{\nabla}\cdot\langle \underline{\underline{\pi}} \rangle_\alpha = \underline{\nabla}\cdot\langle \underline{\underline{\tau}} \rangle_\alpha - \underline{\nabla}\langle P \rangle_\alpha \tag{58}$$

If the continuous fluid phase is Newtonian and isothermal, then the phase average of the stress tensor is

$$\langle \underline{\underline{\tau}} \rangle_\alpha = \mu[\langle \underline{\nabla v} \rangle_\alpha + \langle (\underline{\nabla v})^T \rangle_\alpha] - \frac{2}{3}\mu\langle \underline{\nabla}\cdot\underline{v} \rangle_\alpha \underline{I} \tag{59}$$

where the viscosity is constant due to the isothermal assumption. The phase average, Equation (7), and the averaging theorem, Equation (23), are used to obtain the phase average of the gradient of the velocity, its transpose and the divergence of the velocity in Equation (59). The divergence of Equation (59) gives the viscous force

$$\underline{\nabla}\cdot\langle \underline{\underline{\tau}} \rangle_\alpha = \mu\nabla^2\langle \underline{v} \rangle_\alpha + \mu\underline{\nabla}\cdot(\underline{\nabla}\langle \underline{v} \rangle_\alpha)^T - \frac{2}{3}\mu\underline{\nabla}(\underline{\nabla}\cdot\langle \underline{v} \rangle_\alpha) +$$

$$\mu\underline{\nabla}\cdot[\frac{1}{V_D}\int_{A_{\alpha\beta}}(\underline{v}\underline{n}_{\alpha\beta} + \underline{n}_{\alpha\beta}\underline{v})d_\xi A] - \frac{2}{3}\mu\underline{\nabla}\,[\frac{1}{V_D}\int_{A_{\alpha\beta}}\underline{v}\cdot\underline{n}_{\alpha\beta}d_\xi A] \tag{60}$$

where the integrals over the interfaces are negligible because there is no slip at the boundaries, the **β**-phase velocity is small and does not change very much (ref. 3) with position.

Equation (60) is substituted into Equation (56) to obtain

$$\varepsilon_\alpha\langle\rho\rangle_\alpha^\alpha[\frac{\partial}{\partial t}\bar{\underline{v}}^\alpha + \bar{\underline{v}}^\alpha\cdot\underline{\nabla}\bar{\underline{v}}^\alpha] = \mu\nabla^2\langle \underline{v} \rangle_\alpha + \frac{1}{3}\mu\underline{\nabla}(\underline{\nabla}\cdot\langle \underline{v} \rangle_\alpha) - \underline{\nabla}\varepsilon_\alpha\langle P \rangle_\alpha^\alpha +$$

$$\varepsilon_\alpha\langle\rho\rangle_\alpha^\alpha\bar{\underline{g}}^\alpha - \underline{F} - \underline{\nabla}\cdot(\varepsilon_\alpha\langle\rho\rangle_\alpha^\alpha\overline{\tilde{\underline{v}}^\alpha\tilde{\underline{v}}^\alpha}^\alpha) + \frac{1}{V_D}\int_{A_{\alpha\beta}}\rho\tilde{\underline{v}}^\alpha(\underline{w} - \underline{v})\cdot\underline{n}_{\alpha\beta}d_\xi A \tag{61}$$

where

$$\underline{F} \equiv -\frac{1}{V_D}\int_{A_{\alpha\beta}}\underline{\underline{\pi}}\cdot\underline{n}_{\alpha\beta}d_\xi A \tag{62}$$

is the interfacial drag due to surface forces and appears as a loss of **α**-phase momentum. Equation (61) is the form of the multiphase momentum equation that is restricted to fluid **α**-phases which are Newtonian.

The Permeability Function

The interfacial drag force, given in Equation (62), requires a constitutive equation in which the Darcian permeability appears as the constitutive parameter.

In terms of the viscous and pressure forces, the drag force can be written as

$$\underline{F} = -\frac{1}{V_D}\int_{A_{\alpha\beta}} (\underline{\underline{\tau}} - P\underline{\underline{I}})\cdot\underline{n}_{\alpha\beta} d_{\xi}A \tag{63}$$

which becomes

$$\underline{F} = -\frac{1}{V_D}\int_{A_{\alpha\beta}} \underline{\underline{\tau}}\cdot\underline{n}_{\alpha\beta} d_{\xi}A + \frac{1}{V_D}\int_{A_{\alpha\beta}} P\underline{n}_{\alpha\beta} d_{\xi}A \tag{64}$$

The first term on the right is the viscous drag due to α-phase flow and the second term on the right can be considered as either the form drag or the buoyant force.

The viscous drag term must be proportional to the difference in velocity between the two phases such that

$$-\frac{1}{V_D}\int_{A_{\alpha\beta}} \underline{\underline{\tau}}\cdot\underline{n}_{\alpha\beta} d_{\xi}A = \frac{\varepsilon_{\alpha}^{2}\mu}{K}(\langle\underline{v}\rangle_{\alpha}^{\alpha} - \langle\underline{v}\rangle_{\beta}^{\beta}) \tag{65}$$

where the coefficient of the velocity difference has been obtained by dimensional analysis using the Buckingham Pi theorem (ref. 11) and K is the permeability.

The velocity difference must be between the intrinsic phase average velocities rather than the superficial or phase average velocities to permit the viscous drag to become zero when there is no relative motion between the phases. This velocity difference also makes Equation (65) material frame indifferent (ref. 12).

For the form drag in Equation (64), an area average can be defined by

$$\hat{P}^{\alpha}\int_{A_{\alpha\beta}} \underline{n}_{\alpha\beta} d_{\xi}A = \int_{A_{\alpha\beta}} P\underline{n}_{\alpha\beta} d_{\xi}A \tag{66}$$

For closed channel flow, the pressure changes only in the direction of flow and is uniform across any cross-section. Thus, the average of the pressure on the interfaces between the two phases is the same as the intrinsic phase average pressure over the volume of the α-phase, that is

$$\hat{P}^{\alpha} = \langle P\rangle_{\alpha}^{\alpha} \tag{67}$$

The combination of Equations (25), (66), and (67) gives

$$-\langle P\rangle_{\alpha}^{\alpha}\underline{\nabla}\varepsilon_{\alpha} = \frac{1}{V_D}\int_{A_{\alpha\beta}} P\underline{n}_{\alpha\beta} d_{\xi}A \tag{68}$$

When Equations (65) and (68) are substituted into Equation (64), the expression for the drag becomes

$$-\underline{F} = -\frac{\varepsilon_\alpha^2 \mu}{K}(\langle \underline{v}\rangle_\alpha^\alpha - \langle \underline{v}\rangle_\beta^\beta) + \langle P\rangle_\alpha^\alpha \underline{\nabla}\varepsilon_\alpha \tag{69}$$

and the multiphase momentum equation with constitutive equations for the Newtonian viscous force and interphase drag force is

$$\varepsilon_\alpha \langle\rho\rangle_\alpha^\alpha [\frac{\partial}{\partial t}\bar{\underline{v}}^\alpha + \bar{\underline{v}}^\alpha \cdot \underline{\nabla}\bar{\underline{v}}^\alpha] = \mu\nabla^2\langle \underline{v}\rangle_\alpha + \frac{1}{3}\mu\underline{\nabla}(\underline{\nabla}\cdot\langle \underline{v}\rangle_\alpha) - \varepsilon_\alpha \underline{\nabla}\langle P\rangle_\alpha^\alpha + \varepsilon_\alpha \langle\rho\rangle_\alpha^\alpha \bar{\underline{g}}^\alpha -$$

$$\frac{\varepsilon_\alpha^2 \mu}{K}(\langle \underline{v}\rangle_\alpha^\alpha - \langle \underline{v}\rangle_\beta^\beta) - \underline{\nabla}\cdot\varepsilon_\alpha\langle\rho\rangle_\alpha^\alpha \overline{\tilde{\underline{v}}^\alpha \tilde{\underline{v}}^\alpha}^\alpha + \frac{1}{V_D}\int_{A_{\alpha\beta}} \rho\tilde{\underline{v}}^\alpha(\underline{w} - \underline{v})\cdot\underline{n}_{\alpha\beta} d_\xi A \tag{70}$$

Hookean Elastic Particulate Phase

The β-phase equation of momentum is analogous to Equation (56) and its specific form is

$$\varepsilon_\beta \langle\rho\rangle_\beta^\beta [\frac{\partial}{\partial t}\bar{\underline{v}}^\beta + \bar{\underline{v}}^\beta \cdot \underline{\nabla}\bar{\underline{v}}^\beta] = \underline{\nabla}\cdot\langle\underline{\underline{\pi}}\rangle_\beta + \varepsilon_\beta \langle\rho\rangle_\beta^\beta \bar{\underline{g}}^\beta - \underline{\nabla}\cdot\varepsilon_\beta\langle\rho\rangle_\beta^\beta \overline{\tilde{\underline{v}}^\beta \tilde{\underline{v}}^\beta}^\beta +$$

$$\frac{1}{V_D}\int_{A_{\beta\alpha}} \underline{\underline{\pi}}\cdot\underline{n}_{\beta\alpha} d_\xi A + \frac{1}{V_D}\int_{A_{\beta\alpha}} \rho\tilde{\underline{v}}^\beta(\underline{w} - \underline{v})\cdot\underline{n}_{\beta\alpha} d_\xi A \tag{71}$$

If the β-phase is composed of solid particulates that are deformable such that each particle is an isotropic, elastic solid following Hooke's law, then

$$\underline{\underline{\pi}} = \lambda\phi\underline{\underline{I}} + G\underline{\underline{\gamma}} \tag{72}$$

where λ is the Lame constant and G is the shear modulus. The dilatation of the β-phase, ϕ, is defined by

$$\phi = \frac{1}{2}(\gamma_{11} + \gamma_{22} + \gamma_{33}) \tag{73}$$

which is also the first and, for an isotropic β-phase, the only invariant of the strain tensor. The components of the strain tensor are

$$\gamma_{ij} = \frac{\partial u_i}{\partial x_j} + \frac{\partial u_j}{\partial x_i} \tag{74}$$

and the u_i are the components of the strain vector, $\underline{u}$. The expression for the stress force in Equation (71) becomes

$$\underline{\nabla}\cdot\langle\underline{\underline{\pi}}\rangle_\beta = \underline{\nabla}\varepsilon_\beta\bar{\lambda}^\beta\bar{\phi}^\beta + \underline{\nabla}\langle\tilde{\lambda}^\beta\tilde{\phi}^\beta\rangle_\beta + G\underline{\nabla}\cdot\langle\underline{\underline{\gamma}}\rangle_\beta \tag{75}$$

Before this expression is inserted into Equation (71), the utility of the jump momentum balance, Equation (55), can be demonstrated by using it to find an alternate expression for the **β**-phase drag force.

When the jump momentum balance, Equation (55), and the definition of the drag force, Equation (62), are used in the expression for the **β**-phase drag force, the result is

$$\frac{1}{V_D}\int_{A_{\beta\alpha}} \underline{\underline{\pi}}\cdot\underline{n}_{\beta\alpha}d_\xi A = \underline{F} - \frac{1}{V_D}\int_{A_{\alpha\beta}} \rho\underline{v}(\underline{w} - \underline{v})\cdot\underline{n}_{\alpha\beta}d_\xi A - \frac{1}{V_D}\int_{A_{\beta\alpha}} \rho\underline{v}(\underline{w} - \underline{v})\cdot\underline{n}_{\beta\alpha}d_\xi A \quad (76)$$

Substitution of Equations (75) and (76) into Equation (71) gives

$$\varepsilon_\beta\langle\rho\rangle_\beta^\beta\left[\frac{\partial}{\partial t}\bar{\underline{v}}^\beta + \bar{\underline{v}}^\beta\cdot\underline{\nabla}\bar{\underline{v}}^\beta\right] = \underline{\nabla}\varepsilon_\beta\bar{\lambda}^\beta\bar{\phi}^\beta + \underline{\nabla}\langle\tilde{\lambda}^\beta\tilde{\phi}^\beta\rangle_\beta + G\underline{\nabla}\cdot\langle\underline{\underline{\gamma}}\rangle_\beta + \varepsilon_\beta\langle\rho\rangle_\beta^\beta\bar{\underline{g}}^\beta +$$

$$\underline{F} - \underline{\nabla}\cdot\varepsilon_\beta\langle\rho\rangle_\beta^\beta\overline{\tilde{\underline{v}}^\beta\tilde{\underline{v}}^\beta}^\beta + \frac{1}{V_D}\int_{A_{\alpha\beta}} \rho\tilde{\underline{v}}^\beta(\underline{w} - \underline{v})\cdot\underline{n}_{\beta\alpha}d_\xi A -$$

$$\frac{1}{V_D}\int_{A_{\alpha\beta}} \left[\rho\underline{v}(\underline{w} - \underline{v})\cdot\underline{n}_{\beta\alpha}|_\beta + \rho\underline{v}(\underline{w} - \underline{v})\cdot\underline{n}_{\alpha\beta}|_\alpha\right]d_\xi A \quad (77)$$

where the drag force, $\underline{F}$, is given by Equation (69).

This is the **β**-phase momentum equation for a deformable, Hookean elastic, particulate **β**-phase and, when compared with the **α**-phase momentum equation, Equation (61), the different signs on the drag force indicate that the momentum lost by the fluid phase is gained by the solid phase.

Non-Deformable Particulate Phase

It is unlikely for the flows encountered in filtration that the drag forces are sufficient to cause most solid particulates phases to deform. Hence, consideration of non-deformable particulates should encompass a signifigant number of slurries.

The **β**-phase drag term in Equation (71) can be written as

$$\int_{A_{\beta\alpha}} \underline{\underline{\pi}}\cdot\underline{n}_{\beta\alpha}d_\xi A = \hat{\underline{\underline{\pi}}}^\beta\cdot\int_{A_{\beta\alpha}} \underline{n}_{\beta\alpha}d_\xi A \quad (78)$$

where $\hat{\underline{\underline{\pi}}}^\beta$ is the average stress over the interfacial surface between the two phases. Since

$$\underline{n}_{\alpha\beta} = -\underline{n}_{\beta\alpha} \quad (79)$$

and

$$\underline{\nabla}\varepsilon_\alpha = -\underline{\nabla}\varepsilon_\beta \quad (80)$$

then Equation (25) can be written as

$$\underline{\nabla}\varepsilon_\beta = -\frac{1}{V_D}\int_{A_{\beta\alpha}} \underline{n}_{\beta\alpha}d_\xi A \quad (81)$$

In a non-deformable phase, the stress exists only on the surface of the particulate and according to Whitaker (ref. 13)

$$\underline{\underline{\hat{\pi}}}^{\beta} = \langle\underline{\underline{\pi}}\rangle_{\beta}^{\beta} \tag{82}$$

The combination of Equations (78), (81), and (82) gives

$$\langle\underline{\underline{\pi}}\rangle_{\beta}^{\beta}\cdot\underline{\nabla}\varepsilon_{\beta} = -\frac{1}{V_D}\int_{A_{\alpha\beta}}\underline{\underline{\pi}}\cdot\underline{n}_{\beta\alpha}d_{\xi}A \tag{83}$$

which, when substituted into Equation (71), gives

$$\varepsilon_{\beta}\langle\rho\rangle_{\beta}^{\beta}[\frac{\partial}{\partial t}\underline{\bar{v}}^{\beta} + \underline{\bar{v}}^{\beta}\cdot\underline{\nabla}\underline{\bar{v}}^{\beta}] = \varepsilon_{\beta}\underline{\nabla}\cdot\langle\underline{\underline{\pi}}\rangle_{\beta}^{\beta} + \varepsilon_{\beta}\langle\rho\rangle_{\beta}^{\beta}\underline{\bar{g}}^{\beta} - \underline{\nabla}\cdot\varepsilon_{\beta}\langle\rho\rangle_{\beta}^{\beta}\overline{\underline{\tilde{v}}^{\beta}\underline{\tilde{v}}^{\beta}}^{\beta} +$$

$$\frac{1}{V_D}\int_{A_{\alpha\beta}}\rho\underline{\tilde{v}}^{\beta}(\underline{w} - \underline{v})\cdot\underline{n}_{\beta\alpha}d_{\xi}A \tag{84}$$

Equation (84) is the multiphase momentum equation for a non-deformable particulate phase.

Fickian Diffusion in Both Phases

As described earlier, each phase is assumed to contain two chemical species. Then in each phase, the following equations (ref. 9)

$$\underline{j}_1 = -\rho D_{12}\underline{\nabla}\omega_1 \tag{85}$$

$$\underline{j}_2 = -\rho D_{21}\underline{\nabla}\omega_2 \tag{86}$$

$$\underline{j}_1 = -\underline{j}_2 \tag{87}$$

$$\underline{\nabla}\omega_1 = -\underline{\nabla}\omega_2 \tag{88}$$

$$D_{12} = D_{21} \tag{89}$$

hold.

The phase average of Fick's law is

$$\langle\underline{j}_i\rangle_{\alpha} = -\rho D_{ij}\langle\underline{\nabla}\omega_i\rangle_{\alpha} \quad , \; i = 1,2 \; , \; j \neq i \tag{90}$$

where the phase density, Equation (44), and mass diffusivity are constant if the solution is dilute. This equation can be converted to

$$\langle\underline{j}_i\rangle_{\alpha} = -\rho D_{ij}[\underline{\nabla}\varepsilon_{\alpha}\langle\omega_i\rangle_{\alpha}^{\alpha} + \frac{1}{V_D}\int_{A_{\alpha\beta}}\omega_i\underline{n}_{\alpha\beta}d_{\xi}A] \quad , \; i = 1,2 \; , \; j \neq i \tag{91}$$

which can then be written as

$$\langle\underline{j}_i\rangle_{\alpha} = -\rho D_{ij}[\varepsilon_{\alpha}\underline{\nabla}\langle\omega_i\rangle_{\alpha}^{\alpha} + \frac{1}{V_D}\int_{A_{\alpha\beta}}(\omega_i - \langle\omega_i\rangle_{\alpha}^{\alpha})\underline{n}_{\alpha\beta}d_{\xi}A] \quad , \; i = 1,2 \; , \; j \neq i \tag{92}$$

Since the density is constant, then

$$\langle\omega_i\rangle_\alpha^\alpha = \bar{\omega}_i^\alpha \quad , \; i = 1,2 \tag{93}$$

and the phase average of the mass flux is

$$\langle \underline{j}_i \rangle_\alpha = -\rho D_{ij}\left[\varepsilon_\alpha \nabla \bar{\omega}_i^\alpha + \frac{1}{V_D}\int_{A_{\alpha\beta}} \tilde{\omega}_i^\alpha \underline{n}_{\alpha\beta} d_\xi A\right] \quad , \; i = 1,2 \; , \; j \neq i \tag{94}$$

The divergence of the phase average of the diffusion flux is

$$\underline{\nabla}\cdot\langle \underline{j}_i \rangle_\alpha = -\rho D_{ij}\underline{\nabla}\cdot(\varepsilon_\alpha \underline{\nabla}\bar{\omega}_i^\alpha) - \rho D_{ij}\underline{\nabla}\cdot\left[\frac{1}{V_D}\int_{A_{\alpha\beta}} \tilde{\omega}_i^\alpha \underline{n}_{\alpha\beta} d_\xi A\right] \quad , \; i = 1,2 \; , \; j \neq i \tag{95}$$

which can be substituted in the multiphase species balance, Equation (48), to obtain

$$\frac{\partial}{\partial t}\langle\rho\rangle_\alpha \bar{\omega}_i^\alpha + \underline{\nabla}\cdot\langle\rho\rangle_\alpha \bar{\omega}_i^\alpha \underline{\bar{v}}^\alpha - \underline{\nabla}\cdot\varepsilon_\alpha\left[\langle\rho\rangle_\alpha^\alpha\left(D_{ij}\underline{\nabla}\bar{\omega}_i^\alpha + \overline{\tilde{\omega}_i^\alpha \underline{\tilde{v}}^\alpha}^\alpha\right) + \underline{I}_\omega\right] +$$

$$\frac{1}{V_D}\int_{A_{\alpha\beta}} \rho\omega_i(\underline{v} - \underline{w})\cdot\underline{n}_{\alpha\beta} d_\xi A + \frac{1}{V_D}\int_{A_{\alpha\beta}} \underline{j}_i\cdot\underline{n}_{\alpha\beta} d_\xi A = \langle r_i\rangle_\alpha \quad , \; i = 1,2 \; , \; j \neq i \tag{96}$$

where

$$\varepsilon_\alpha \underline{I}_\omega \equiv \frac{1}{V_D}\int_{A_{\alpha\beta}} \tilde{\omega}_i^\alpha \underline{n}_{\alpha\beta} d_\xi A \quad , \; i = 1,2 \tag{97}$$

The continuity condition, Equation (37), can be combined with the first two terms in Equation (96) to give

$$\frac{\partial}{\partial t}\varepsilon_\alpha\langle\rho\rangle_\alpha^\alpha \bar{\omega}_i^\alpha + \underline{\nabla}\cdot\varepsilon_\alpha\langle\rho\rangle_\alpha^\alpha \bar{\omega}_i^\alpha \underline{\bar{v}}^\alpha = \varepsilon_\alpha\langle\rho\rangle_\alpha^\alpha\left[\frac{\partial}{\partial t}\bar{\omega}_i^\alpha + \underline{\bar{v}}^\alpha\cdot\underline{\nabla}\bar{\omega}_i^\alpha\right] -$$

$$\omega_i\left[\frac{1}{V_D}\int_{A_{\alpha\beta}} \rho(\underline{v} - \underline{w})\cdot\underline{n}_{\alpha\beta} d_\xi A\right] \quad , \; i = 1,2 \tag{98}$$

If

$$\underline{I}_\omega \doteq 0 \quad , \quad \overline{\tilde{\omega}_i^\alpha \underline{\tilde{v}}^\alpha}^\alpha \doteq 0 \quad , \; i = 1,2 \tag{99}$$

then Equations (98) and (99) can be combined with Equation (96) to give

$$\varepsilon_\alpha\langle\rho\rangle_\alpha^\alpha\left[\frac{\partial}{\partial t}\bar{\omega}_i^\alpha + \underline{\bar{v}}^\alpha\cdot\underline{\nabla}\bar{\omega}_i^\alpha\right] - \langle\rho\rangle_\alpha^\alpha D_{ij}\underline{\nabla}\cdot\varepsilon_\alpha\underline{\nabla}\bar{\omega}_i^\alpha + \frac{1}{V_D}\int_{A_{\alpha\beta}} \rho\omega_i(\underline{v} - \underline{w})\cdot\underline{n}_{\alpha\beta} d_\xi A -$$

$$\omega_i\left[\frac{1}{V_D}\int_{A_{\alpha\beta}} \rho(\underline{v} - \underline{w})\cdot\underline{n}_{\alpha\beta} d_\xi A\right] + \frac{1}{V_D}\int_{A_{\alpha\beta}} \underline{j}_i\cdot\underline{n}_{\alpha\beta} d_\xi A = \langle r_i\rangle_\alpha \quad , \; i = 1,2 \; , \; j \neq i \tag{100}$$

For the solute, or 2 species, the interfacial transfer of mass by diffusion, which is the last term on the left in Equation (100), can be written in terms of a mass transfer coefficient, $k^\bullet_{loc}$,

$$-\frac{1}{V_D}\int_{A_{\alpha\beta}}\underline{j}_2\cdot\underline{n}_{\alpha\beta}d_\xi A = \varepsilon_\alpha a\dot{k}_{loc}(\langle\rho_2\rangle_\alpha^\alpha|_\alpha - \langle\rho_2\rangle_\alpha^\alpha) \tag{101}$$

where the flux is positive into the fluid α-phase and

$$a \equiv \frac{A_{\alpha\beta}}{V_D} \tag{102}$$

The fluid phase solute species equation for dilute solutions in which the tortuosity and dispersion, Equation (99), are negligible is

$$\varepsilon_\alpha\langle\rho\rangle_\alpha^\alpha[\frac{\partial}{\partial t}\bar{\omega}_2^\alpha + \underline{\bar{v}}^\alpha\cdot\underline{\nabla}\bar{\omega}_2^\alpha] = \langle\rho\rangle_\alpha^\alpha D_{21}\underline{\nabla}\cdot\varepsilon_\alpha\underline{\nabla}\bar{\omega}_2^\alpha + \varepsilon_\alpha a\dot{k}_{loc}(\langle\rho_2\rangle_\alpha^\alpha|_\alpha - \langle\rho_2\rangle_\alpha^\alpha) +$$

$$\omega_2[\frac{1}{V_D}\int_{A_{\alpha\beta}}\rho(\underline{v} - \underline{w})\cdot\underline{n}_{\alpha\beta}d_\xi A] - \frac{1}{V_D}\int_{A_{\alpha\beta}}\rho\omega_2(\underline{v} - \underline{w})\cdot\underline{n}_{\alpha\beta}d_\xi A + \langle r_2\rangle_\alpha \tag{103}$$

By interchanging the roles of α and β, analogous equations can be written for the solid particulate β-phase.

THE MULTIPHASE EQUATIONS FOR VARIOUS PARTICULATE CHARACTERISTICS

Slightly Soluble, Non-deformable, Particulate Phase in a Newtonian Fluid

The general multiphase equations can be applied to filter cakes for which specific assumptions are made. In this section, it assumed that the fluid phase is an incompressible Newtonian liquid phase in which there are two chemical species present. One of the species in the liquid phase is the solvent liquid and the solute is the slightly soluble particulate phase. There is no reaction occurring in the liquid phase, the entire system is isothermal, and Fick's law of diffusion is effective in the liquid phase.

The particulate phase is composed of only one species and has no reaction or concentration gradients. This phase is non-deformable but it is slightly soluble in the liquid phase.

For the liquid phase Equations (37), (43), (51), (55), (70), (100), and (101) apply and Equation (84) and the analogous form of Equation (37)

$$\frac{\partial}{\partial t}\varepsilon_\beta\langle\rho\rangle_\beta^\beta + \underline{\nabla}\cdot\varepsilon_\beta\langle\rho\rangle_\beta^\beta\underline{\bar{v}}^\beta + \frac{1}{V_D}\int_{A_{\beta\alpha}}\rho(\underline{v} - \underline{w})\cdot\underline{n}_{\beta\alpha}d_\xi A = 0 \tag{104}$$

for the particulate phase.

For the solvent species, labeled 1, in the liquid phase at the liquid solid interfaces, the jump species balance, Equation (51), is

$$[\rho_1(\underline{v} - \underline{w}) + \underline{j}_1]\cdot\underline{n}_{\alpha\beta}|_\alpha + [\rho_1(\underline{v} - \underline{w}) + \underline{j}_1]\cdot\underline{n}_{\beta\alpha}|_\beta = 0 \tag{105}$$

Since the particulate phase has no concentration gradients

$$\underline{j}_1|_\beta = 0 \tag{106}$$

and

$$\rho_1(\underline{v} - \underline{w})\cdot\underline{n}_{\beta\alpha}|_\beta = 0 \tag{107}$$

because the concentration of the liquid phase in the particulate phase is zero. In addition

$$\underline{j}_1|_\alpha = 0 \tag{108}$$

because the liquid phase is insoluble in the particulate phase. Consequently,

$$\rho_1(\underline{v} - \underline{w})\cdot\underline{n}_{\alpha\beta}|_\alpha = 0 \tag{109}$$

which means that on the liquid side of the interface the liquid velocity and interface velocity are the same.

For the solute species in the liquid phase at the liquid solid interfaces, the jump species balance is

$$[\rho_2(\underline{v} - \underline{w}) + \underline{j}_2]\cdot\underline{n}_{\alpha\beta}|_\alpha + [\rho_2(\underline{v} - \underline{w}) + \underline{j}_2]\cdot\underline{n}_{\beta\alpha}|_\beta = 0 \tag{110}$$

Since there are no concentration gradients in the particulate phase, then

$$\underline{j}_2|_\beta = 0 \tag{111}$$

and

$$\rho_2(\underline{v} - \underline{w})\cdot\underline{n}_{\alpha\beta}|_\alpha = 0 \tag{112}$$

from Equation (109). Equation (110) then reduces to

$$\underline{j}_2\cdot\underline{n}_{\alpha\beta}|_\alpha = -\rho_2(\underline{v} - \underline{w})\cdot\underline{n}_{\beta\alpha}|_\beta \tag{113}$$

where the solute species density in the particulate phase is the density of the particulate phase.

Once the interface conditions have been determined from the system specifications, then the multiphase equations can be simplified.

The multiphase mass equation for the liquid phase, Equation (37), is reduced to

$$\frac{\partial\varepsilon_\alpha}{\partial t} + \underline{\nabla}\cdot\varepsilon_\alpha\underline{\bar{v}}^\alpha = 0 \tag{114}$$

by using Equation (44) for the constant density liquid phase and the interface condition, Equation (109).

The appropriate form of the mass equation for the particulate phase, Equation (104), is obtained by combining the interface condition, Equation (113), with the mass transfer coefficient expression, Equation (101). The result, for a constant density particulate phase, is

$$\rho_\beta\left(\frac{\partial\varepsilon_\beta}{\partial t} + \underline{\nabla}\cdot\varepsilon_\beta\underline{\bar{v}}^\beta\right) = -\varepsilon_\alpha a\dot{k}_{loc}\rho_\alpha(\bar{\omega}_2^\alpha|_\alpha - \bar{\omega}_2^\alpha) \tag{115}$$

which indicates a transfer of the particulate phase to the liquid phase by dissolution.

The solute species equation in the liquid phase, Equation (103), becomes

$$\varepsilon_\alpha[\frac{\partial}{\partial t}\bar{\omega}_2^\alpha + \underline{\bar{v}}^\alpha\cdot\underline{\nabla}\bar{\omega}_2^\alpha] = D_{21}\underline{\nabla}\cdot\varepsilon_\alpha\underline{\nabla}\bar{\omega}_2^\alpha + \varepsilon_\alpha a k_{1oc}^\bullet(\bar{\omega}_2^\alpha|_\alpha - \bar{\omega}_2^\alpha) \quad (116)$$

by combining the interface condition, Equation (109), and the specification that there is no chemical reaction in the liquid phase.

The condition that the liquid phase is not soluble in the particulate phase, Equation (109), reduces the liquid phase momentum equation, Equation (70), to

$$\varepsilon_\alpha\rho_\alpha[\frac{\partial}{\partial t}\underline{\bar{v}}^\alpha + \underline{\bar{v}}^\alpha\cdot\underline{\nabla}\underline{\bar{v}}^\alpha] = \mu\nabla^2\varepsilon_\alpha\underline{\bar{v}}^\alpha + \frac{1}{3}\mu\underline{\nabla}(\underline{\nabla}\cdot\varepsilon_\alpha\underline{\bar{v}}^\alpha) - \varepsilon_\alpha\underline{\nabla}\bar{P}^\alpha + \varepsilon_\alpha\rho_\alpha\underline{\bar{g}}^\alpha - \frac{\varepsilon_\alpha^2\mu}{K}(\underline{\bar{v}}^\alpha - \underline{\bar{v}}^\beta) \quad (117)$$

where the dispersion of momentum

$$\underline{\nabla}\cdot\varepsilon_\alpha\rho_\alpha\overline{\underline{\tilde{v}}^\alpha\underline{\tilde{v}}^\alpha}^\alpha \doteq 0 \quad (118)$$

has been neglected.

The interface condition given in Equation (113) is used in the momentum equation for a non-deformable particulate phase, Equation (84), to obtain

$$\varepsilon_\beta\rho_\beta[\frac{\partial}{\partial t}\underline{\bar{v}}^\beta + \underline{\bar{v}}^\beta\cdot\underline{\nabla}\underline{\bar{v}}^\beta] = \varepsilon_\beta\underline{\nabla}\cdot\underline{\underline{\bar{\pi}}}^\beta + \varepsilon_\beta\rho_\beta\underline{\bar{g}}^\beta - \frac{1}{V_D}\int_{A_{\alpha\beta}}\underline{\tilde{v}}^\beta(\underline{j}_2\cdot\underline{n}_{\alpha\beta})d_\xi A \quad (119)$$

where the particulate phase dispersion of momentum

$$\underline{\nabla}\cdot\varepsilon_\beta\rho_\beta\overline{\underline{\tilde{v}}^\beta\underline{\tilde{v}}^\beta}^\beta \doteq 0 \quad (120)$$

is again neglected. The integral over the interfacial area in Equation (119) indicates that there is an interchange of momentum between the two phases due to the dissolution of the particulate phase.

Equations (114), (115), (116), (117), and (119) describe the filtration of a slightly soluble particulate phase in a Newtonian liquid phase.

Insoluble, Elastic, Deformable, Particulate Phase in a Newtonian Fluid

The multiphase equations are next applied to a constant density, Newtonian liquid phase with only one chemical species and no chemical reaction occurring. The particulate phase is elastic and deformable but has only one chemical species, no reaction, constant density, and is not soluble in the liquid phase.

The surface velocity of the particulate phase is the vector sum of the displacement velocity of the particle and the velocity due to the deformation of the particle. Consequently, the porosity, according to Equation (29), is a function of time.

Due to the specification that neither phase is soluble in the other and that there is no mass transfer between the phases implies that, at the interfaces between the two phases, the surface and phase velocities are the same and the jump mass balance, Equation (43), is identically satisfied.

With the phase and interface velocities equal to each other and the constant density assumption for both phases, the multiphase mass equations, Equations (37) and (104), reduce to

$$\frac{\partial \varepsilon_\alpha}{\partial t} + \underline{\nabla} \cdot \varepsilon_\alpha \underline{\bar{v}}^\alpha = 0 \tag{121}$$

$$\frac{\partial \varepsilon_\beta}{\partial t} + \underline{\nabla} \cdot \varepsilon_\beta \underline{\bar{v}}^\beta = 0 \tag{122}$$

The multiphase momentum equation for the Newtonian fluid phase is Equation (117), where the intergal over the interfaces between the two phases is zero because the liquid and interface velocities are equal and where the dispersion of momentum, Equation (118) is again neglected. The jump momentum balance, Equation (55), reduces to

$$\underline{\underline{\pi}} \cdot \underline{n}_{\alpha\beta}|_\alpha + \underline{\underline{\pi}} \cdot \underline{n}_{\beta\alpha}|_\beta = 0 \tag{123}$$

If the deformable particulate phase is an elastic Hookean solid for which the Lame constant and the dilatation are constant, then, under the restrictions imposed on the particulate phase, the particulate phase momentum equation, Equation (77), becomes

$$\varepsilon_\beta \rho_\beta [\frac{\partial}{\partial t} \underline{\bar{v}}^\beta + \underline{\bar{v}}^\beta \cdot \underline{\nabla \bar{v}}^\beta] = G \underline{\nabla} \cdot \underline{\bar{\gamma}}^\beta - (\lambda \phi + \bar{P}^\alpha) \underline{\nabla} \varepsilon_\alpha + \frac{\varepsilon_\alpha^2 \mu}{K} (\underline{\bar{v}}^\alpha - \underline{\bar{v}}^\beta) + \varepsilon_\beta \rho_\beta \underline{\bar{g}}^\beta \tag{124}$$

where Equation (69) has been used for the interfacial drag force and the dispersion of particulate phase momentum, Equation (120), has again been neglected.

Insoluble, Non-Deformable, Particulate Phase in a Newtonian Liquid

Of the three cases considered, this case in which the particulate phase is non-deformable is probably the one which is most likely to be encountered in filtration practice. The fluid phase is again a Newtonian liquid and each phase has a single species, is incompressible and is without chemical reaction.

Since the velocities of both phases at the interface are same as the interface velocity, then the jump mass balance, Equation (43), is again satisfied identically. The mass equations for both phases again reduce to those given by Equations (121) and (122) and the liquid phase momentum balance is given by Equation (117). The particulate phase momentum equation, Equation (84), reduces to

$$\varepsilon_\beta \rho_\beta [\frac{\partial}{\partial t} \underline{\bar{v}}^\beta + \underline{\bar{v}}^\beta \cdot \underline{\nabla \bar{v}}^\beta] = \varepsilon_\beta \underline{\nabla} \cdot \underline{\underline{\bar{\pi}}}^\beta + \varepsilon_\beta \rho_\beta \underline{\bar{g}}^\beta \tag{125}$$

ONE-DIMENSIONAL FILTRATION OF NON-DEFORMABLE PARTICULATES

The theortical development provides general multiphase equations for the filtration of soluble, elastic, and non-deformable particulates suspended in a Newtonian incompressible liquid.

The procedure for developing a specific filtration equation for the filtration of a non-deformable particulatae phase in a Newtonian liquid is described next. This procedure can also be extended to the soluble and elastic particulate phases.

Dimensional Analysis and Dominant Effects

The filtration of a non-deformable particulate phase from a Newtonian liquid phase is described in general by Equations (121), (122), (117), and (125). In the liquid phase momentum equation, Equation (117), there are five forces; the inertial force, the viscous force, the pressure force, the gravity force, and the interfacial drag force, and in the particulate phase momentum equation, Equation (125), there are three forces; the inertial force, the stress force and the gravity force. To determine which of the forces in each of these equations are dominant, dimensional analysis is used.

Dimensional analysis is an estimation procedure that involves the selection of characteristic quantities that can be measured and that make both the dimensionless independent and dependent variables range between zero and unity or, if not unity, a value of the order of unity.

Once these characteristic quantities have been selected, the terms in the dimensionless equations then appear as the product of a dimensionless number and a spatial or time derivative. The procedure then assumes that all the derivatives are of the order of unity and that the relative importance of the term is determined by the magnitude of the dimensionless coefficient.

When the dominant terms have been selected, then the assumption is made that an exact solution to the approximate differential equation is also an approximate solution the exact differential equation. Segal (ref.14) points out that this latter assumption is not generally true. However, for the determination of the filtrate rate it turns out that the approximate differential equation does not need to be solved but only evaluated at a boundary.

The dimensionless forms of Equations (121), (122), (117) and (125) are

$$\frac{\partial \varepsilon_\alpha}{\partial t^*} + [Sr]\underline{\nabla}^* \cdot \varepsilon_\alpha \underline{v}_\alpha^* = 0 \tag{126}$$

$$\frac{\partial \varepsilon_\beta}{\partial t^*} + [Sr]\underline{\nabla}^* \cdot \varepsilon_\beta \underline{v}_\beta^* = 0 \tag{127}$$

$$\left[\frac{Re_\alpha \varepsilon_\alpha}{Sr}\right]\frac{\partial}{\partial t^*}\underline{v}_\alpha^* + \left[\frac{Re_\alpha \varepsilon_\alpha}{Sr}\right]\underline{v}_\alpha^* \cdot \underline{\nabla}^* \underline{v}_\alpha^* = \nabla^{*2} \varepsilon_\alpha \underline{v}_\alpha^* + \frac{1}{3}\underline{\nabla}^*(\underline{\nabla}^* \cdot \varepsilon_\alpha \underline{v}_\alpha^*) -$$

$$[Np_\alpha]\varepsilon_\alpha \underline{\nabla}^* P_\alpha^* - [Nd_\alpha]\varepsilon_\alpha(\underline{v}_\alpha^* - \underline{v}_\beta^*) + \left[\frac{Re_\alpha}{Fr_\alpha}\right]\varepsilon_\alpha \underline{g}_\alpha^* \tag{128}$$

$$\left[\frac{Nr_{\beta}\varepsilon_{\beta}}{Sr}\right]\frac{\partial}{\partial t^{*}}\underline{v}_{\beta}^{*} + \left[\frac{Nr_{\beta}\varepsilon_{\beta}}{Sr}\right]\underline{v}_{\beta}^{*}\cdot\underline{\nabla}^{*}\underline{v}_{\beta}^{*} = \varepsilon_{\beta}\underline{\nabla}^{*}\cdot\underline{\underline{\pi}}_{\beta}^{*} - \left[\frac{Nr_{\beta}}{Fr_{\beta}}\right]\varepsilon_{\beta}\underline{g}_{\beta}^{*} \tag{129}$$

where the definitions and physical significance of the dimensionless groups are given in Table 1.

TABLE 1

Definitions and physical significance of the dimensionless groups that appear in the dimensionless multiphase equations of change.

Symbol	Name	Expression	Physical Significance
Sr	Strouhal	1	Convective acceleration/local acceleration.
Re_{α}	Reynolds	$\rho_{\alpha}\dot{V}_{F}L/A\mu\varepsilon_{\alpha}^{*}$	α-phase inertial force/α-phase viscous force.
Np_{α}	-	$\dot{P}_{c}LA\varepsilon_{\alpha}^{*}/\mu\dot{V}_{F}$	α-phase pressure force/α-phase viscous force.
Nd_{α}	-	$\varepsilon_{\alpha}^{*}L^{2}/K_{o}$	Interphase drag force/α-phase viscous force.
$Re_{\alpha}Fr_{\alpha}^{-1}$	Reynolds/Froude	$\rho_{\alpha}gL^{2}A\varepsilon_{\alpha}^{*}/\mu\dot{V}_{F}$	α-phase gravity force/α-phase viscous force.
Nr_{β}	-	$\rho_{\beta}\dot{V}_{F}^{2}/\varepsilon_{\alpha}^{*2}A^{2}\Pi_{o}$	β-phase inertial force/β-phase deformation force.
$Nr_{\beta}Fr_{\beta}^{-1}$	-	$\rho_{\beta}gL/\Pi_{o}$	β-phase gravity force/β-phase deformation force.

The expressions for the dimensionless coefficients shown in Table 1 indicate that the cake length, the septum permeability, and the stress experienced by the septum are new quantities that must be measured in addition to measuring the usual filtration quantities such as liquid and particulate properties, filter area, filtrate rate, and cake pressure drop.

The experimental apparatus for making such measurements is shown in Fig. 3. Photographs of the transparent filter chamber, on which a millimeter scale is etched, permit timed recordings of the cake length. Axial pressure distributions are measured with pressure transducers connected to liquid filled probes located at .01, 0.5, 1.0, 2.0, 3.0, 4.0, and 6.0 centimeters above the septum. The pressure drop across the cake is controlled by venting the pressure applied to the slurry tank. Two calibrated rotameters with matched pressure drops are arranged in parallel to measure exit flow rates. The filtrate volume, the stress experienced by the septum and the slurry concentrations in the recycle loop about the slurry tank are continuously monitored.

All of the variables listed in Table 1 can be measured directly except the permeability at the cake septum interface, K_o. To calculate this quantity, it is necessary to make the hypothesis that the pressure, gravitational, and drag forces are the dominant forces in Equation (117). If the calculated dimensionless numbers contradict this hypothesis, then additional terms would have to be considered in the calculation of K_o. However, the results shown in Table 2 confirm this hypothesis.

The numerical values of the dimensionless numbers shown in Table 2 indicate that in the liquid phase, the inertial and viscous forces are negligible and that the pressure, gravitational, and interfacial drag forces are the dominant forces. In the particulate phase, the dominant forces are the deformation and gravitational forces.

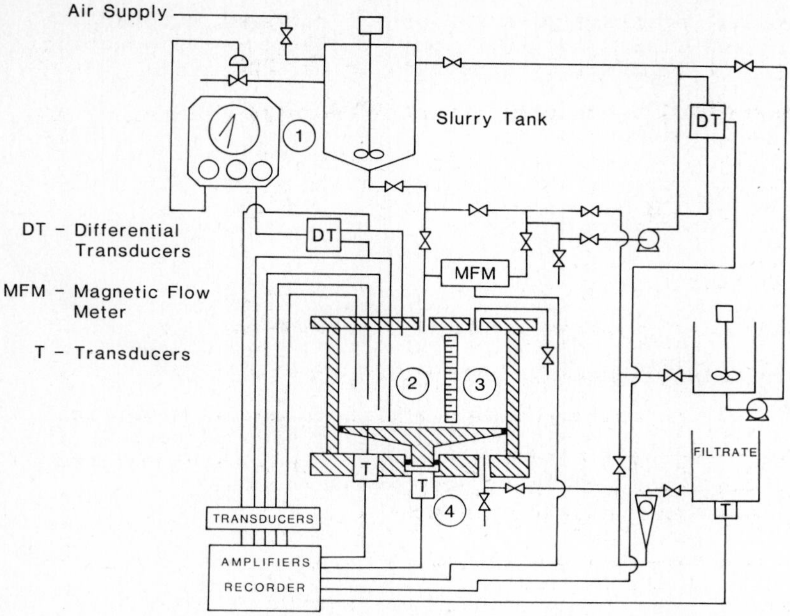

1. Pressure controller maintains a constant cake pressure drop.
2. Pressure probes measure the pressure distribution.
3. Scale on Plexiglas filter chamber measures cake length.
4. Movable septum measures septum stress.

Fig. 3. A schematic of the filtration apparatus.

TABLE 2

Filtration data and corresponding values of the dimensionless coefficients.

ρ_α = 997. kgm^{-3} $\quad$ μ = $9.69 \times 10^{-4} Nsm^{-2}$ $\quad$ s = .101

ρ_β = 1180. kgm^{-3} $\quad$ A = .0324 m^2 $\quad$ ε_α^* = .393

$P_c^{\cdot} 10^{-4}$ Nm^{-2}	$\dot{V}_F 10^4$ m^3s^{-1}	L m	$K_o 10^{12}$ m^2	$\Pi_o 10^{-4}$ Nm^{-2}	Re_α	$\frac{Re_\alpha}{Fr_\alpha}$	Np_α	Nd_α	Nr_β	$\frac{Nr_\beta}{Fr_\beta}$
5.85	1.57	.0355	2.49	8.04	450	1.0×10^6	170×10^6	200×10^6	1.97×10^{-6}	.0051
7.49	1.30	.0540	3.11	9.50	570	2.9×10^6	410×10^6	370×10^6	1.30×10^{-6}	.0066
7.04	0.95	.0715	3.01	10.3	550	7.0×10^6	700×10^6	700×10^6	$.63 \times 10^{-6}$	.0080
6.98	0.69	.0916	3.16	9.56	540	17×10^6	1300×10^6	1100×10^6	$.36 \times 10^{-6}$	.011
6.99	0.56	.119	3.14	9.39	540	32×10^6	1900×10^6	1800×10^6	$.25 \times 10^{-6}$	.015
6.83	0.42	.154	3.04	8.75	530	72×10^6	3000×10^6	3100×10^6	$.18 \times 10^{-6}$	.020

The dimensional analysis does not affect the mass equations, Equations (121) and (122), but the liquid phase momentum equation, Equation (117), for filtration is reduced to

$$K\underline{\nabla}\bar{P}^{\alpha} = -\varepsilon_{\alpha}\mu(\underline{\bar{v}}^{\alpha} - \underline{\bar{v}}^{\beta}) + \rho_{\alpha}\underline{\bar{g}}^{\alpha} \tag{130}$$

and the particulate phase momentum equation, Equation (125), becomes

$$\underline{\nabla}\cdot\underline{\bar{\pi}}^{\beta} = -\rho\underline{\bar{g}}^{\beta} \tag{131}$$

One-Dimensional Filtration

Most empirical filtration studies are performed in a vertical, cylindrical geometry with one-dimensional cake accumulation in the axial direction. The simplified multiphase equations, Equations (121), (122), (130), and (131), in one dimension are

$$\frac{\partial\varepsilon_{\alpha}}{\partial t} = \frac{\partial\langle v\rangle_{\alpha z}}{\partial z} \tag{132}$$

$$\frac{\partial\varepsilon_{\beta}}{\partial t} = \frac{\partial\langle v\rangle_{\beta z}}{\partial z} \tag{133}$$

$$K\frac{\partial\bar{P}^{\alpha}}{\partial z} = \varepsilon_{\alpha}\mu(\bar{v}_z^{\alpha} - \bar{v}_z^{\beta}) - \rho_{\alpha}g \tag{134}$$

$$\frac{\partial}{\partial z}\bar{\pi}_{zz}^{\beta} = \rho_{\beta}g \tag{135}$$

where the origin is taken at the cake septum interface and the positive direction is taken upward such that

$$\underline{\bar{v}}^{\alpha} = -\bar{v}^{\alpha}\underline{k} \tag{136}$$

$$\underline{\bar{v}}^{\beta} = -\bar{v}^{\beta}\underline{k} \tag{137}$$

$$\underline{\bar{g}}^{\alpha} = \underline{\bar{g}}^{\beta} = -g\underline{k} \tag{138}$$

Equation (9) has also been used for the velocities in the mass equations for both phases.

Examination of the mass equations for both phases, Equations (132) and (133), indicates that they are not independent. If the distribution of the particulate phase is known through the porosity function, then the velocities of both phases are specified. This occurs because the viscous forces are negligible which implies that the velocity profile is flat in the filter cake and a function of only axial position. Consequently, the fluid phase velocity depends only on the cross sectional area available for flow, that is the porosity, and the continuity condition.

Time appears explicitly in the mass equations for both phases, Equations (132) and (133), but not in the momentum equations, Equations (134) and (135), because the inertial forces in the momentum equations for both phases are negligible for filtrations. This means that the momentum equations hold at every instant throughout the unsteady-state filtration. If the mass and momentum equations are to be connected, then time must be implicitly introduced into the momentum equations by the change of independent variable from z to ξ, that is

$$\xi = \frac{z}{L(t)} \tag{139}$$

It is also convenient to combine the gravity and pressure terms in the momentum equations for both phases. The measurable liquid phase pressure, relative to the pressure at the cake septum interface, is the sum of the flow pressure and the hydrostatic pressure and is defined by

$$\dot{P}_\alpha \equiv \bar{P}^\alpha + \rho_\alpha g \xi L(t) - P_0 \tag{140}$$

The stress in the particulate phase due only to flow and relative to the stress experienced by the septum is given by

$$\dot{\pi}_{\beta zz} = \bar{\pi}^\beta_{zz} - \rho_\beta g \xi L(t) - \pi_0 \tag{141}$$

Both of the quantities defined in Equations (140) and (141) can be made dimensionless by dividing by the measurable pressure drop across the cake

$$\dot{P}_c = P + \rho_\alpha g L - P_0 \tag{142}$$

With these changes, the mass and momentum equations for a one-dimensional filtration, Equations (132), (133), (134), and (135) become

$$L\frac{\partial \varepsilon_\alpha}{\partial t} = \xi \dot{L}\frac{\partial \varepsilon_\alpha}{\partial \xi} + \frac{\partial \langle v \rangle_{\alpha z}}{\partial \xi} \tag{143}$$

$$L\frac{\partial \varepsilon_\beta}{\partial t} = \xi \dot{L}\frac{\partial \varepsilon_\beta}{\partial \xi} + \frac{\partial \langle v \rangle_{\beta z}}{\partial \xi} \tag{144}$$

$$\dot{P}_c\frac{\partial}{\partial \xi}\dot{P}^*_\alpha = \frac{\varepsilon_\alpha \mu L}{K}(\bar{v}^\alpha_z - \bar{v}^\beta_z) \tag{145}$$

$$\dot{P}_c\frac{\partial}{\partial \xi}\dot{\pi}^*_{\beta zz} = 0 \tag{146}$$

The particulate momentum equation, Equation (146), indicates that, for a non-deformable particulate phase, the stress is uniform at any instant during the filtration. This means that a non-deformable particulate phase behaves as a simple "conductor" of momentum from the liquid phase to the particulate phase and finally to the septum. This is the extent of the usefulness of Equation (146).

This uniform particulate phase stress is not the same as the cumulative drag stress, which is discussed by Willis and Tosun (ref. 3). As shown previously, if the particulate phase is elastic and deformable then Hooke's law can be used for the particulate phase stress. This extension to deformable particulates cannot be made with the cumulative drag concept.

Functionality of the Local Porosity

Further simplification of the mass equations can be made if the functionality of the local porosity can be determined. For a one-dimensional filtration, the local porosity can be a function of only time, or a function of fractional position and time, or a function of only fractional position.

If the porosity is a function of only time then the mass equations, Equations (143) and (144), can be integrated to obtain the axial velocity profiles

$$\langle v \rangle_{\alpha z} = \frac{\dot{V}_F}{A} - L(t)\left|\frac{d\varepsilon_\alpha}{dt}\right|\xi \tag{147}$$

$$\langle v \rangle_{\beta z} = L(t)\left|\frac{d\varepsilon_\alpha}{dt}\right|\xi \tag{148}$$

where it has been assumed that the porosity can only decrease with time, that is

$$-\frac{d\varepsilon_\alpha}{dt} = \left|\frac{d\varepsilon_\alpha}{dt}\right| \tag{149}$$

and the boundary conditions at the cake septum interface

$$\langle v \rangle_{\alpha z}\big|_{\xi=0} = \dot{V}_F A^{-1} \tag{150}$$

$$\langle v \rangle_{\beta z}\big|_{\xi=0} = 0 \tag{151}$$

have been used. These velocity profiles are too restrictive to be admissible and the possibility that the local porosity is a function of only time is rejected.

It is possible to distinguish between the two remaining alternatives by determining the average porosity

$$\varepsilon_\alpha^* \equiv L^{-1}\int_0^L \varepsilon_\alpha(z,t)dz = \int_0^1 \varepsilon_\alpha(\xi,t)d\xi \tag{152}$$

during the course of the filtration. If the average porosity is constant, then the local porosity is a function of only fractional cake position but if the average porosity is a function of time, then the local porosity is a function of fractional position and time.

Indirect, Continuous Measurement of the Average Porosity

An indirect, but continuous, indication of the average porosity during the course of the filtration can be obtained from the external mass balance

$$\frac{M_\beta}{s} - \rho_\alpha V_F = M_\alpha + M_\beta \tag{153}$$

by eliminating the mass of the liquid and particulates in the cake with

$$M_\alpha = \rho_\alpha \varepsilon^*_\alpha AL \tag{154}$$

$$M_\beta = \rho_\beta \varepsilon^*_\beta AL \tag{155}$$

to obtain a linear relation

$$L = GV_F \tag{156}$$

between the cake length and filtrate volume if the slope

$$G = \frac{s\rho_\alpha}{A[\rho_\beta \varepsilon^*_\alpha(1 - s) - \rho_\alpha \varepsilon^*_\alpha s]} \tag{157}$$

is constant.

Consequently, if the cake length and filtrate volume are measured during the course of a filtration in which the slurry concentration is constant and the cake length is linear in filtrate volume, then the average porosity is constant and the local porosity is a function of only fractional cake position. If the cake length is not linear in filtrate volume, then the local porosity is a function of both fractional cake position and time.

Fig. 4 shows the cake length as a function of filtrate volume for filtrations performed in the apparatus shown in Fig. 3 of Lucite in water at slurry concentrations of 10.1%, 7.49%, and 5.27% and cake pressure drops of 68 kPa, 120 kPa, and 71 kPa, respectively. The linearity that is exhibited indicates that the average porosity is constant throughout the filtration and that the local porosity is a function of only fractional cake position.

The only other explicit data on cake length and filtrate volume is that reported by Shirato, et al. (ref. 15) for Hong Kong pink kaolin and this data also shows the same lineariy that is shown in Fig. 4. It can be deduced then that Hong Kong pink kaolin gives constant average porosity filter cakes in which the local porosity is a function of only fractional cake position.

Equation (153) is the difference between the slurry filtered from the slurry tank and the amount of filtrate collected and is independent of the fluid, the particulate phase characteristics, and the geometry of the filter cake or filter chamber. Consequently, the cake volume is linear with filtrate volume for any

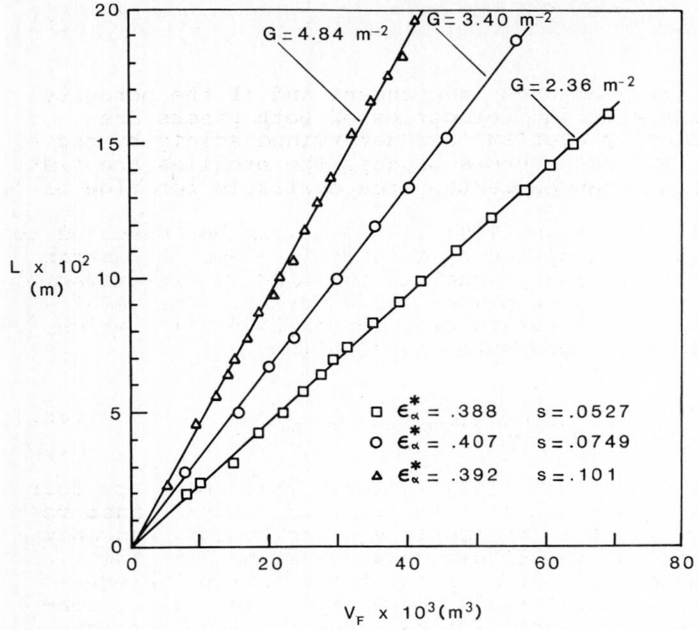

Fig. 4. Cake length as a function of filtrate volume for filtrations of Lucite in water at three levels of slurry concentration.

geometry filter chamber and any fluid-particulate phase combination if the slurry concentration and average porosity are constant.

Constitutive Equations, Experiments, and Filter Design

On the basis that the average porosity is constant and the local porosity is a function of only fractional position, then Equations (143) and (144) become

$$-\xi\dot{L}\frac{d\varepsilon_\alpha}{d\xi} = \frac{\partial\langle v\rangle_{\alpha z}}{\partial\xi} \tag{158}$$

$$-\xi\dot{L}\frac{d\varepsilon_\beta}{d\xi} = \frac{\partial\langle v\rangle_{\beta z}}{\partial\xi} \tag{159}$$

These two equations can be integrated to obtain

$$\langle v\rangle_{\alpha z} = \dot{V}_F A^{-1} - \dot{L}\int_0^\xi \xi\frac{d\varepsilon_\alpha}{d\xi}d\xi \tag{160}$$

$$\langle v\rangle_{\beta z} = \dot{L}\int_0^{\xi}\xi\frac{d\varepsilon_\alpha}{d\xi}d\xi \tag{161}$$

Equations (160) and (161) are not independent and if the porosity distribution is given, then the velocities of both phases are specified. These velocity profiles are determined solely by the continuity condition for both phases because the profiles are flat and respond only to the changes in the area available for flow or the porosity.

The velocities in Equations (160) and (161) can be converted to intrinsic phase average velocities by Equation (9) and the intrinsic phase average velocities are equal to the mass average velocities when the densities of both phases are constant. The resulting velocities can then be substituted into Equation (145) which, upon integration, gives the pressure distribution

$$P_\alpha^{\cdot *}(\xi,t) = \frac{\mu G\dot{V}_F V_F}{A\dot{P}_c}\left[\int_0^{\xi}\frac{d\xi}{K(\xi,t)} - GA\int_0^{\xi}\frac{(1 + \varepsilon_\alpha/\varepsilon_\beta)}{K(\xi,t)}\left[\int_0^{\xi}\eta\frac{d\varepsilon_\alpha}{d\eta}d\eta\right]d\xi\right] \tag{162}$$

Equations (160), (161), and (162) indicate that there are four unknown functions; the pressure distribution, the liquid phase velocity, the porosity, and the permeability distribution, and only two independent equations; the liquid phase mass and momentum equations. As a consequence, even if the two constitutive equations for the particulate and liquid phases are given (i.e., nondeformable and Newtonian), two additional constitutive relations are required, one for the distribution of the two phases, or the porosity, and one for the pore geometry, or permeability, before the internal velocity and pressure distributions can be determined.

Predictive filter design requires solutions to Equations (160) and (162) and these solutions need either constitutive relations for the porosity and permeability or experimental data to supply the equivalent informtion if these two constitutive relations are not available. This equivalent experimental information is usually the pressure and porosity distributions since these two are easier to measure than the permeability or velocity distributions.

Determination of Internal Cake Variations

The procedure for determining the internal permeability and velocity distributions consists of measuring the local distribution of pressure (ref. 15) and the local distribution of the porosity (ref. 16). A function can be fitted to the measured porosity distribution and then the velocitiy profiles can be obtained from Equations (160) and (161).

The velocity distributions are then combined with the derivative of the fitted function representing the measured pressure distribution in Equation (145) to obtain the permability distribution. An example of this procedure is given by Willis and Tosun (ref. 3).

Liquid Phase Mass Balance for the Filter Cake

A liquid phase mass balance for the entire filter cake can be obtained by evaluating the integral in Equation (160) using integration by parts and then multiplying by the liquid phase density to obtain

$$\rho_\alpha \dot{V}_F = \rho_\alpha A\langle v\rangle_{\alpha z,L} + \frac{d(\rho_\alpha AL)}{dt}(\varepsilon_{\alpha,L} - \overset{*}{\varepsilon}_\alpha) \tag{163}$$

Equation (163) indicates that, within the confines of the filter cake and at any instant, the mass of liquid leaving the cake is equal to the mass of liquid entering the cake plus a mass of liquid that is displaced from the filter cake. The displaced fluid is proportional to the difference between the surface and average porosities and to the cake deposition rate. This amount is small and experimental attempts to measure it have not been successful.

Development of the Filtrate Rate Equation

The filtrate rate equation is obtained by evaluating Equation (145) at the exit of the filter cake

$$\dot{P}_C\left(\frac{\partial}{\partial\xi}\dot{P}_\alpha^{*}\right)\Big|_{\xi=0} = \mu L\left(\frac{\varepsilon_\alpha \bar{v}_z^\alpha}{K}\right)\Big|_{\xi=0} \tag{164}$$

where the velocity of the particulate phase, Equation (151), is zero. Since the intrinsic phase average velocity is equal to the mass average velocity when the liquid phase density is constant, then Equation (9) can be used obtain the superficial velocity which, at the exit of the cake, is given by Equation (150). Equation (164) becomes

$$\dot{P}_C A(K_0 J_0) = \mu L \dot{V}_F \tag{165}$$

where J_o is the dimensionless pressure gradient and K_o is the cake septum permeability, both of which are evaluated at the exit of the filter cake. Combining Equations (156) and (165) with Equation (162) shows that the internal pressure distribution is influenced by the septum permeability.

The exit of the filter cake is located at the point inside the septum where the particulate velocity is zero. If the fibers of the septum are also non-deformable, then the multiphase filtration theory for a non-deformable particulate phase cannot distinguish between the particulates and the septum and this theory then applies up to the cake septum interface where the particulate velocity is zero.

Since the liquid phase velocity is determined by the porosity distribution and liquid phase continuity, Equation (160), then the pressure gradient at the septum must adjust to changes in the septum permeability to maintain the exit flow rate specified by the liquid phase continuity.

The cake length can be eliminated from Equation (165) with Equation (156) to obtain the multiphase filtration equation in linearized form

$$\dot{V}_F^{-1} = \frac{\mu}{A}\left[\frac{G}{K_0(V_F)J_0(V_F)\dot{P}_C(V_F)}\right]V_F \tag{166}$$

where the filtrate volume replaces time as the independent variable. The notation in Equation (166) indicates that the slope of the cake length versus filtrate volume is constant but that the septum permeability, the septum pressure gradient, and the cake pressure drop are all functions of the filtrate volume.

If the reciprocal rate is linear in filtrate volume then the filtrate volume versus time is parabolic. All of the parameters in the slope of the reciprocal rate versus filtrate volume must be constant for parabolic behavior. If these parameters are not constant initially, but at a later time in the filtration they become constant, then the extrapolation of the linear portion of the reciprocal rate must pass through the origin.

Experimental Verification of the Filtrate Rate Equation

Experimental verification of Equation (166) requires the measurement of the filtrate rate, the filtrate volume, and the pressure drop across the cake. If parabolic behavior is desired, then some means must be used to attain a constant pressure drop across the cake for some part of the filtration. With these measured quantities, together with the liquid viscosity and filter area, the product of the cake septum permeability and cake septum pressure gradient can be obtained.

To separate this product and determine the cake septum permeability, it is necessary to determine experimentally the pressure gradient at the cake septum interface. The measured pressure distribution can be fitted with an appropriate curve and then the curve can be differentiated to obtain the pressure gradient at the cake septum interface. This procedure has better reproducibility than direct graphical differentiation of the pressure distribution. The filtration apparatus shown in Fig. 3 is designed to provide the necessary data to accomplish these calculations.

The most significant parameter in the verification of this multiphase theory of filtration is the cake septum permeability. To obtain significant differences in cake septum permeability, two different types of filter media are selected. One is a very fine porosity paper felt mat which does not permit a significant penetration by the Lucite particles and the other is a very course porosity felt mat that does permit some penetration by the Lucite particles.

Fig. 5 shows the data for cake length, septum permeability and septum pressure gradient for the filtration of a 10.1% Lucite in water slurry at a nominal cake pressure drop of 69 kPa using a fine porosity Whatman No. 3 filter paper.

The data in Fig. 5 indicates that the cake length is linear in filtrate volume which means that the average porosity is constant since the slurry concentration is constant. The permeability and the pressure gradient at the septum are both constant which means that the Lucite particles are not penetrating or clogging the fine porosity Whatman No. 3 filter paper.

The reciprocal rate data shown in Fig. 6 for the fine porosity non-clogging Whatman No. 3 filter paper indicates that parabolic behavior occurs except for the initial portion during which the the cake pressure drop is rising to its controlled value. Extrapolation of the linear portion passes through the origin as predicted by Equation (166). The intercept of the curved portion of the reciprocal rate data is simply the initial reciprocal rate and is not related to the septum permeability. The septum permeability affects the slope and not the intercept of recipocal rate data.

In another filtration, shown in Fig. 7, all the operating variables, except the filter paper, are identical to those used to collect the data shown in Fig. 5 and Fig. 6. The data shown in Fig. 7 on cake length versus filtrate volume is linear and the

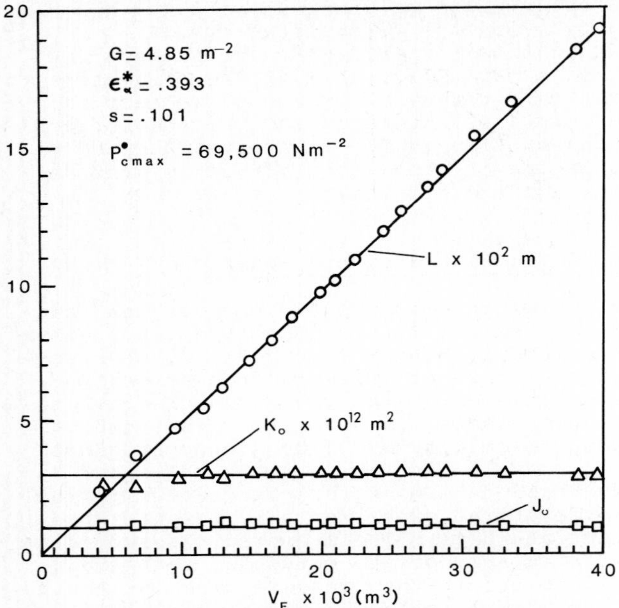

Fig. 5. Cake length, septum permeability and pressure gradient for Lucite in water using a non-clogging Whatman No.3 filter paper.

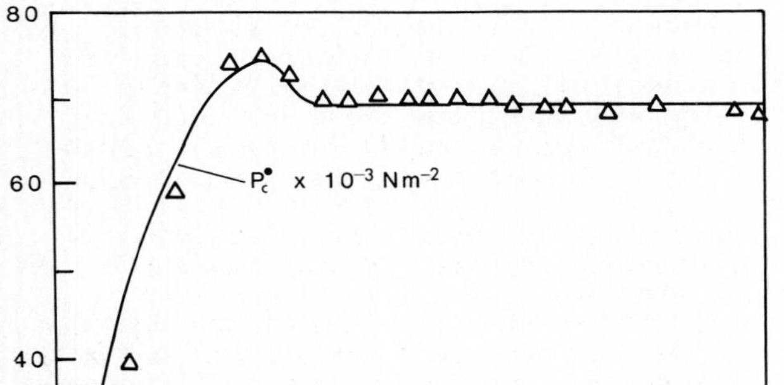

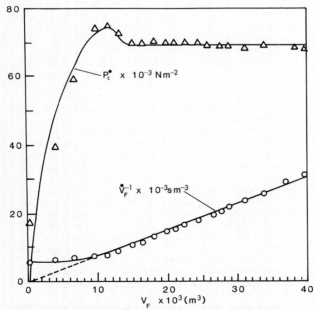

Fig. 6. Cake pressure drop and reciprocal rate for Lucite in water using a non-clogging Whatman No.3 filter paper.

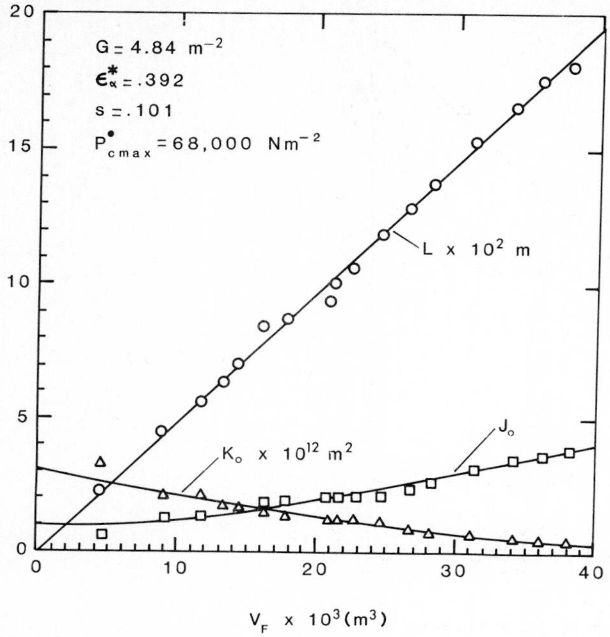

Fig. 7. Cake length, septum permeability and pressure gradient for Lucite in water using a clogging Whatman No. 4 filter paper.

slope and average porosity are the same as those for the fine porosity, non-clogging Whatman No. 3 filter paper that is shown in Fig. 5.

However, the septum permeability for this coarse porosity Whatman No. 4 filter paper decreases and the septum pressure gradient increases which indicates that the filter paper is clogging (see Equation (162)). The clogging is severe enough that at the end of the filtration the pressure gradient at the septum is four times that for the non-clogging septum even when the pressure drop across the entire cake per unit length in both filtrations is the same.

Fig. 8 shows the reciprocal rate data for the clogging Whatman No. 4 filter paper. This data is curved throughout the filtration indicating non-parabolic behavior occurs in the initial stages due to the rapid increase in the cake pressure drop and in the latter stages due to the clogging of the Whatman No. 4 filter paper.

Fig. 9 is a plot of the product of the septum permeability and the septum pressure gradient for the non-clogging and clogging septa and it shows that this product is constant for the Whatman No. 3 non-clogging filter paper and decreases linearly for the Whatman No. 4 filter paper.

These experiments verify that the cake septum permeability, the pressure gradient at the septum, and the cake pressure drop all affect the slope of the reciprocal rate data as predicted by the multiphase filtration theory for a non-deformable particulate phase

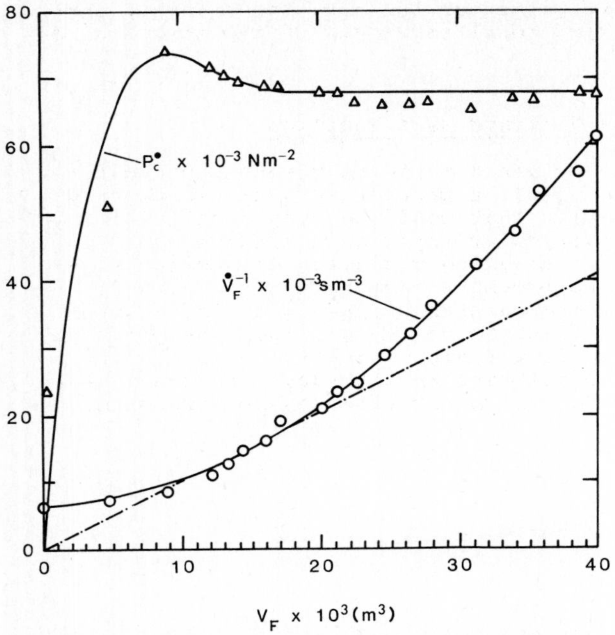

Fig. 8. Cake pressure drop and reciprocal rate for Lucite in water using a clogging Whatman No. 4 filter paper.

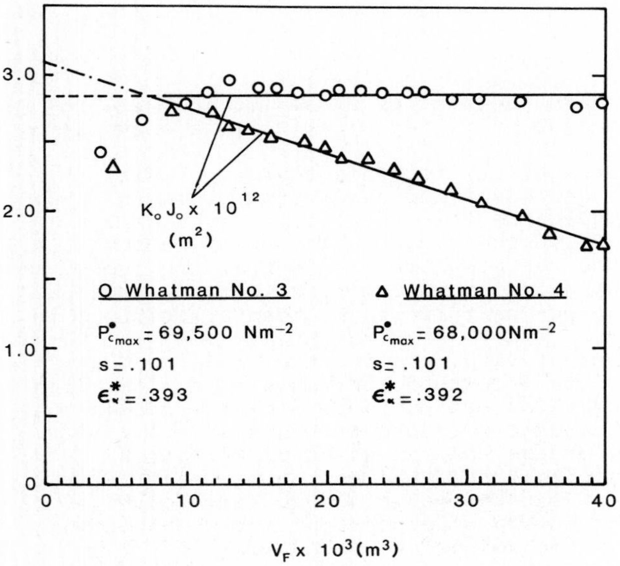

Fig. 9. The product K_oJ_o for a non-clogging Whatman No. 3 filter paper and a clogging Whatman No. 4 filter paper.

in a Newtonian liquid. The parameter that has the greatest influence on the slope of the reciprocal rate data is the selection of the filter medium.

Parametric Analysis of the Filtrate Rate Equation

Since the experimental evidence supports the multiphase filtration equation, Equation (166), a parametric analysis of this equation can yield information that would be very difficult to obtain experimentally. Of particular importance are the effects of the cake pressure drop and a clogging medium on parabolic behavior.

During a filtration, the pressure drop across the cake increases from zero to a maximum constant value in a finite amount of time. Initially, both the filtrate volume and the cake pressure drop are zero but the ratio is a finite, non-zero value that is proportional to the initial filtrate rate through the clean septum. A cake pressure drop function which satisfies these constraints is

$$\dot{P_c} = \dot{P}_{cmax}\left(1 - e^{-(nV_F + m(nV_F)^{\ell})}\right) \tag{167}$$

where n is given by

$$n = \frac{\mu G V_{Fo}}{A(K_oJ_o)_o \dot{P}_{cmax}} = \frac{1}{\dot{P}_{cmax}} \frac{d\dot{P}_c}{dV_F}\bigg|_{V_F=0} \tag{168}$$

and the two adjustable parameters are m and ℓ.

The effect of the cake pressure drop is most evident when the product of the septum permeability and septum pressure gradient is constant. The filtration with the non-clogging Whatman No. 3 filter paper that is shown in Fig. 5, Fig. 6, and Fig. 9 provides the following data

$$(K_oJ_o)_o = K_oJ_o = 2.81 \times 10^{-12}\ m^2 \tag{169}$$

$$\mu = 9.7 \times 10^{-4}\ Nsm^{-2} \tag{170}$$

$$n = 129.6\ m^{-3} \tag{171}$$

$$m = 1.0\ ,\ \ell = 2.0 \tag{172}$$

where the value of m and ℓ fit Equation (167) to the experimental values of cake pressure drop.

The results shown in Fig. 10 are for three values of the adjustable parameter m and show that as the cake pressure drop reaches its constant value more rapidly, the corresponding reciprocal rate exhibits a larger dip from the fixed initial reciprocal rate. The minimum of the dip approaches the origin as the cake pressure drop approaches a step change. A linear reciprocal rate, or parabolic behavior, is achieved as soon as the cake pressure drop becomes constant since the product of the septum pressure gradient and the septum permeability is already constant.

The area under the reciprocal rate versus filtrate volume is the time of the filtration,

$$t = \int_0^{V_F} \dot{V}_F^{-1}\, dV_F \tag{173}$$

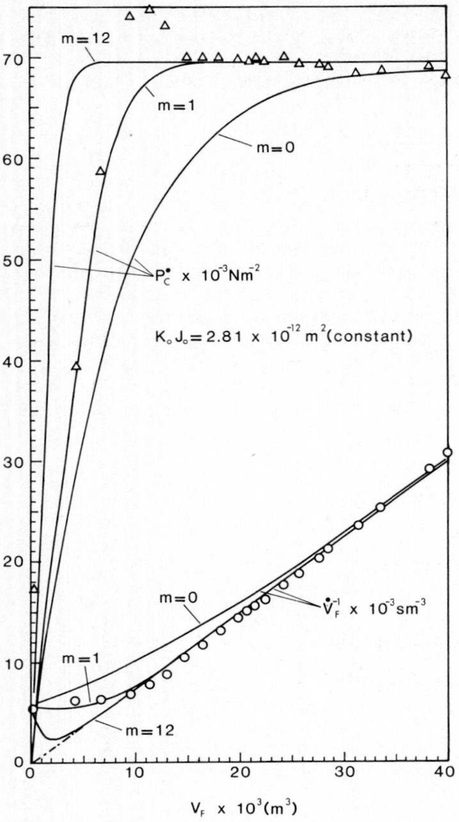

Fig. 10. The effect of cake pressure drop on reciprocal rate.

For a step change in the cake pressure drop, the time of filtration is a minimum. As the rise to the fixed pressure occurs more slowly, the time of filtration increases.

The effect of septum clogging on parabolic behavior is determined from the same data as that used to determined the effect of the cake pressure drop on reciprocal rate. The cake pressure drop function, Equation (167), is fitted to the experimental values using Equations (171) and (172) and then it is postulated that the product of the septum pressure gradient and septum permeability either continously decreases in a linear fashion or decreases linearly to a constant value, as shown in Fig. 11a.

The results of the effect of septum clogging on reciprocal rate is shown Fig. 11b. The effect of a linear decrease of the product $K_o J_o$ to one-half its initial value is identified by the label 1. The initial curvature of the reciprocal rate is influenced more by the rate of increase of pressure than by the decrease in $K_o J_o$. However, once the cake pressure drop becomes constant, the reciprocal rate continues to exhibit curvature, or non-parabolic behavior, due to the continuous clogging of the septum.

If the product $K_o J_o$ decreases linearly to a constant value, as indicated by the label 2 in Fig. 11a and Fig. 11b, then the reciprocal rate exhibits curvature until the product $K_o J_o$ becomes constant and the corresponding reciprocal rate becomes linear. Ex-

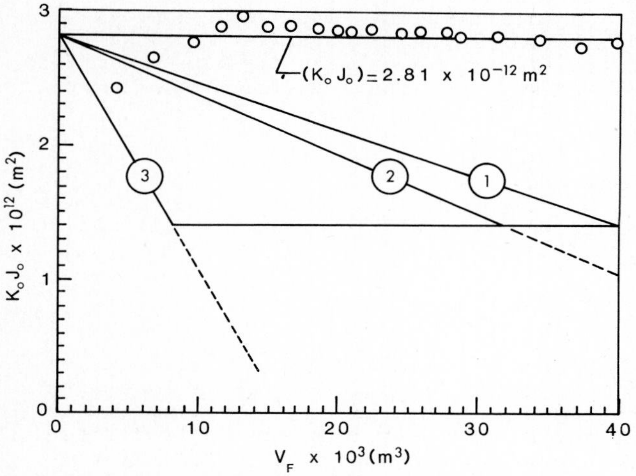

Fig.11a. Three parametric modes of the product $K_o J_o$ for the analysis of the effect of septum clogging on reciprocal rate shown in Fig. 11b

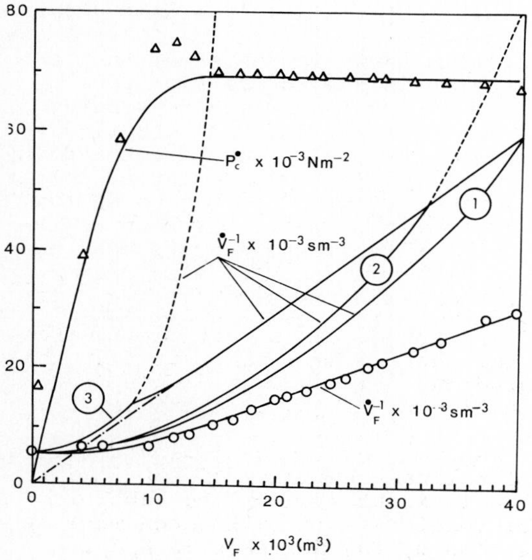

Fig. 11b. The effect of septum clogging on reciprocal filtrate rate.

trapolation of this linear reciprocal rate passes through the origin. If the septum were to continue to clog, then the broken line indicates the path of the reciprocal rate.

If the clogging occurs very rapidly, as indicated by the label 3 in Fig. 11a and Fig. 11b, then curvature in the reciprocal rate is observed again, but, in this case, when the product K_oJ_o becomes constant, curvature is still observed because the cake pressure drop has not yet reached a constant value. Once both effects become constant, parabolic behavior is again observed with the extrapolation of the linear portion again passing through the origin. In all three cases, clogging increases the area under the reciprocal rate curve and hence the filtration time.

Energy Dissipation in Filtration

The energy dissipated by the separation process of filtration can be obtained by calculating the area

$$W = \int_0^{V_F} P dV_F \tag{174}$$

under the pressure versus filtrate volume curve.

Three pressure drops were measured in these filtrations with the clogging and non-clogging septa, the pressure drop across the septum, the pressure drop across the cake, and the applied pressure. The apparatus shown in Fig. 3 is designed to control the pressure drop across the cake at a constant value by venting the applied pressure in the slurry tank.

The applied pressure, the cake pressure drop, and the septum pressure drop for the non-clogging Whatman No. 3 septum and the clogging Whatman No. 4 septum are shown in Fig. 12 and Fig. 13, respectively. Table 3 shows the amount of work and power dissipated through the filter cake, the septum, and the sum of the two for both septa. The power is obtained by dividing the area under the pressure curves by the total time of the filtration. Since the pressure drop across the cake for both septa is the same, then the difference in total energy dissipation is due to the two types of septa. The results show that the non-clogging septum dissipates 46% more total work and 133% more total power than the clogging septum. Consequently, parabolic behavior, which can be achieved by using a non-clogging septum, is significantly less energy efficient than the non-parabolic behavior associated with a clogging septum.

This comparison of energy and power dissipation is made on the basis of the same total volume of filtrate, the same cake mass and average porosity, the same slurry concentration, the same cake pressure drop and the same filter area, and therefore reflects the constraints encountered in the operation of a given filter for a particular slurry. The strategy of operation, then, consists of selecting a septum that dissipates the least amount of energy and yet maintains a prescribed filtrate clarity.

On the other hand, filter design can be considered if the time of filtration is constrained. For example, if the time of filtration for the clogging septum were required to be the same as that for the non-clogging septum, then the filter area for the clogging septum would have to be about 1.6 times that for the non-clogging septum. Consequently, the selection of the septum affects both the capital and operating costs of a filter.

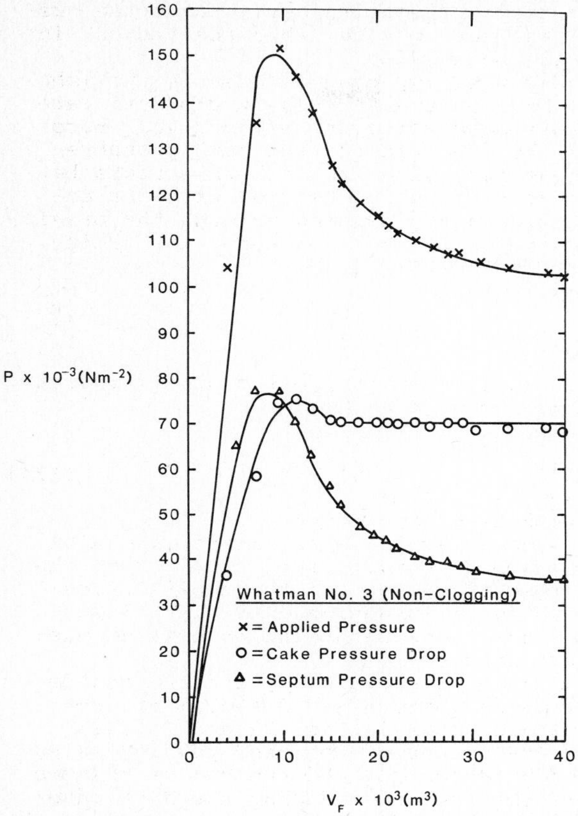

Fig. 12. The applied, cake, and septum pressures for a non-clogging Whatman No. 3 filter paper.

TABLE 3

Dissipation of energy and power between the septum and filter cake for two types of septa.

	Energy Dissipated (Nm) Clogging	Non-Clogging	Increase %
Filter Cake	2580	2550	-
Septum	483	1920	298
Total	3063	4470	46
	Power Dissipated (Nms^{-1}) Clogging (t = 1055s)	Non-Clogging (t = 660s)	Increase %
Filter Cake	2.45	3.86	58
Septum	.46	2.91	533
Total	2.91	6.77	133

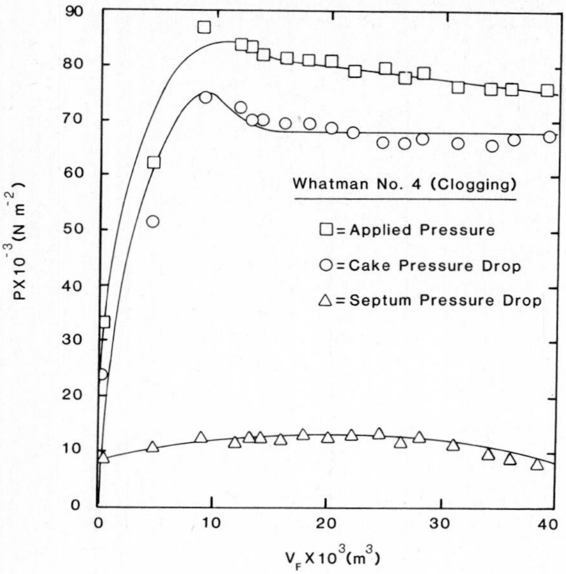

Fig. 13. The applied, cake, and septum pressures for a clogging Whatman No. 4 filter paper.

EFFECT OF OPERATING VARIABLES ON SEPTUM PERMEABILITY

The development and experimental verification of the multiphase filtration theory emphasizes the importance of the septum permeability in determining filtration rates. The significance of the septum has been observed before. Hatschek (ref. 17) was perhaps the first to recognize the influence of the medium permeability on eventual cake structure and to illustrate the effect of extremely thin layers of deposit on the surface of the septum. Ruth, et al. (ref. 18) also observed that the predominant resistance to flow occurs in the layer of particulates adjoining the medium and that restriction of medium pore openings and not added lengths of cake cause the decrease in flow rate. Although their data show that the septum resistance clearly influences the slope of the reciprocal rate versus filtrate volume, the empirical series resistance model forced them to alter the intercept to change the slope.

More recently, Rushton and Griffiths (ref. 19) report that a variable medium resisance is related to curvature in the traditional reciprocal rate plot. Other studies by Grace (ref. 20), Heertjes (ref. 21,22), Kehat, et al. (ref. 23), and Bitter (ref. 24) show that pressure, slurry concentration, media selection, particle size, and particle size distribution affect septum permeability. Of these, only the first three are practical operating variables.

This medium study determines the affect of pressure, slurry concentration, and media selection on the septum permeability, as reflected in K_o, and cake structure, as reflected in the average porosity. The effect of particle size and particle size distribution is minimized by selecting Lucite as the particulate phase since Lucite is composed of nearly uniform, spherical, non-deformable particulates of about 90 microns in diameter.

Statistical Analysis and the Null Hypothesis

The statistical design used in these experiments is a standard (ref. 25,26) 2^n factorial design in which each of the three factors (n=3) is specified at two levels. Two types of septa (Whatman No. 3 and No. 4), two levels of slurry concentration (nominally 5% and 10%), and two levels of pressure (nominally 34 kPa and 103 kPa) are used. Several replications are made at each central point (7.5% and 69 kPa) for each of the two media to obtain error estimates. The measured responses to the three operating factors are the cake septum permeability and the average porosity.

In the analysis of variance of this factorial design, the null hypothesis is that there is no influence from the operating factors on the response variance for the cake septum permeability or the average porosity and that all sample variances estimate the same population variance. The significance of this hypothesis is tested with the single-tailed F-distribution.

Experimental Data

The data for this statistical study are obtained from the filtration apparatus shown in Fig. 3 and the calculation procedure for determining the cake septum permeability has already been described.

The average porosity is calculated from the slope of the cake length versus filtrate volume, Equation (157), and the specified slurry concentration. Average porosities are also determined at the conclusion of each run by successive weighing of the wet and dry filter cake. Comparison of the two porosity values indicates that, within experimental error, the average porosity during the run is the same as that measured at the end of the run.

Typical data is shown in Fig. 14 and Fig. 15. The notable difference between this data is the behavior of the cake septum permeability. In Fig. 14, the Whatman No. 3 paper gives a septum permeability that remains constant throughout the filtration. However, the Whatman No.4 filter paper shown in Fig. 15 has a septum permeability which is initially constant but after a filtrate volume of about .025 m , or at about 2.5 minutes into a 16 minute filtration, the septum permeability decreases until it reaches a value at the end of the filtration that is about one-third of its initial value.

In Fig. 15 the reciprocal rate dips below the linear extrapolation of the data due to the overshoot of the cake pressure drop but once the cake pressure drop becomes constant, the reciprocal rate becomes linear and the filtration with the clogging Whatman No. 4 filter paper exhibits parabolic behavior.

In Fig. 8, the same filter paper is used as that shown in Fig. 15 but in the former the septum permeability decreased more rapidly than the septum pressure gradient increased resulting in a linear decrease in the product K_0J_0 and non-parabolic behavior. Conversly, in Fig. 15 the decrease in the septum permeability is balanced by the increase in the septum pressure gradient which keeps the product K_0J_0 constant and the filtration exhibits parabolic behavior.

Consequently, parabolic behavior alone cannot be used to distinguish between a clogging and a non-clogging septum. It is necessary to measure the internal pressure distribution and calculate the septum permeability before this distinction can be made.

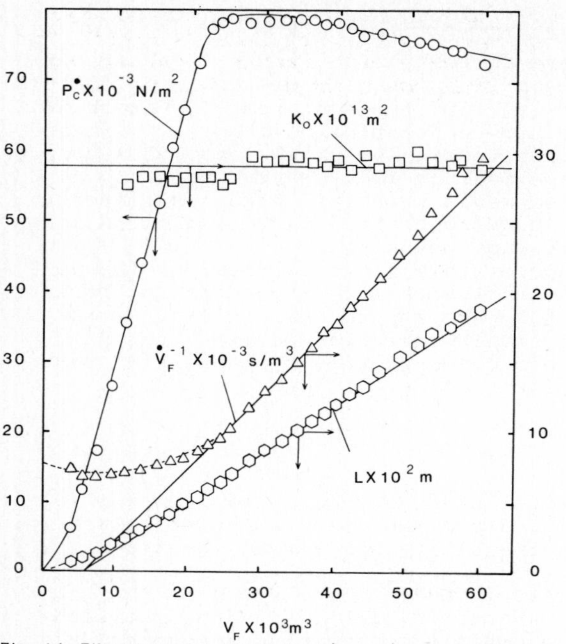

Fig. 14. Filtration of a 7.49% Lucite-water slurry: Whatman No. 3 filter paper, ϵ_{α}^{*} = 0.382, G = 3.50m^{-2}.

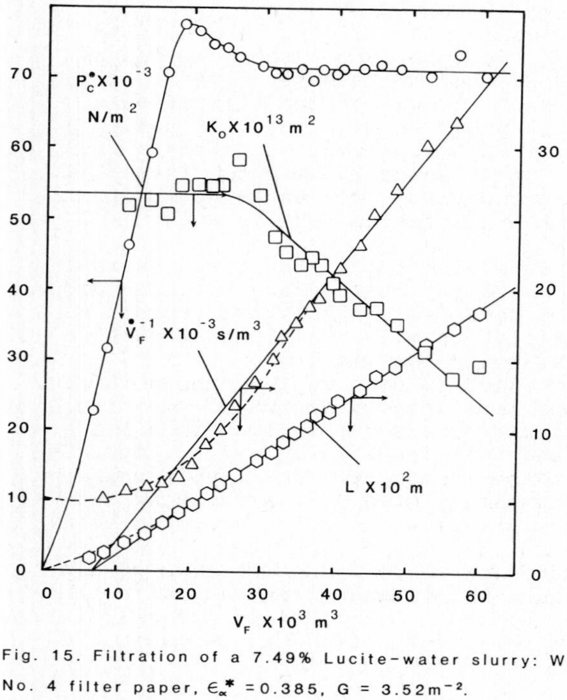

Fig. 15. Filtration of a 7.49% Lucite-water slurry: Whatman No. 4 filter paper, ϵ_{α}^{*} =0.385, G = 3.52m^{-2}.

The pressure profiles in Fig. 16 for the non-clogging Whatman No. 3 filter paper show that the dimensionless pressure profile is linear with fractional position and that the pressure gradient at the septum is constant at unity. This means that

$$\frac{\partial \dot{P}_\alpha}{\partial z}\bigg|_{z=0} = \frac{\dot{P}_c}{L} \tag{175}$$

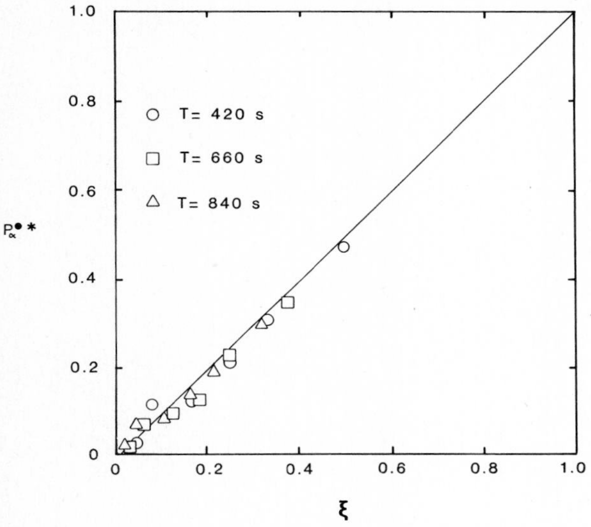

Fig. 16. Axial pressure profiles in a Lucite-water filter cake using a non-clogging Whatman No.3 filter paper.

In Fig. 17, the dimensionless pressure profiles for the clogging Whatman No. 4 filter paper are non-linear with an increasing slope at the cake septum interface. This means that

$$\frac{\partial \dot{P}_\alpha}{\partial z}\bigg|_{z=0} > \frac{\dot{P}_c}{L} \tag{176}$$

which supports the prediction of Equations (162) and (165) that the septum permeability affects the axial pressure distribution.

The cake septum permeabilities as a function of filtrate volume throughout each filtration for all the filtration experiments are shown in Fig. 18 and Fig. 19.

The septum permeabilities shown in Fig. 18 for the non-clogging Whatman No. 3 filter paper indicate no trend and are deduced to be constant.

The septum permeabilities for the clogging Whatman No. 4 filter paper shown in Fig. 19, however, indicate an initial range that is constant followed by a definite, nearly linear, decreasing trend.

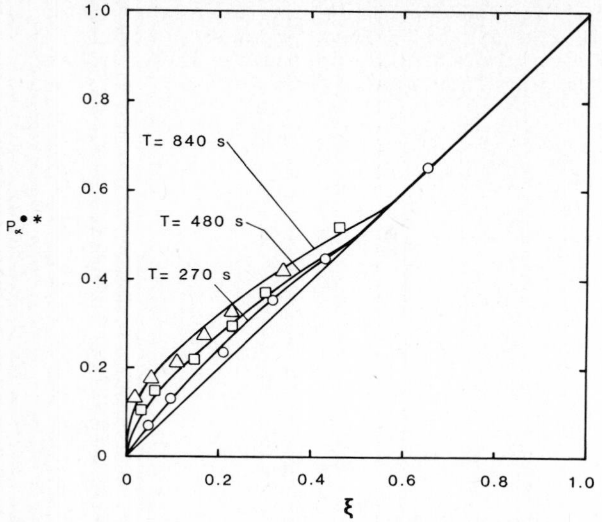

Fig. 17. Axial pressure profiles in a Lucite-water filter cake using a clogging Whatman No. 4 filter paper.

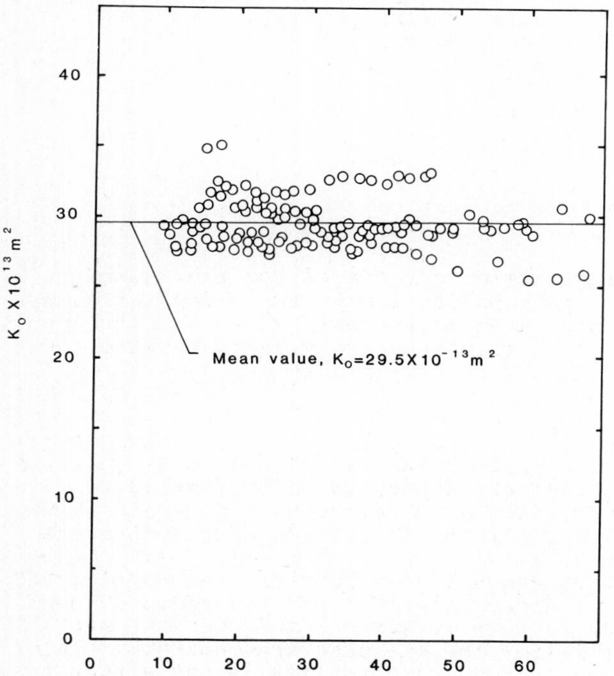

Fig. 18. Cake septum permeabilities throughout each filtration for all the filtration experiments using a Whatman No. 3 filter paper in the statistical analysis.

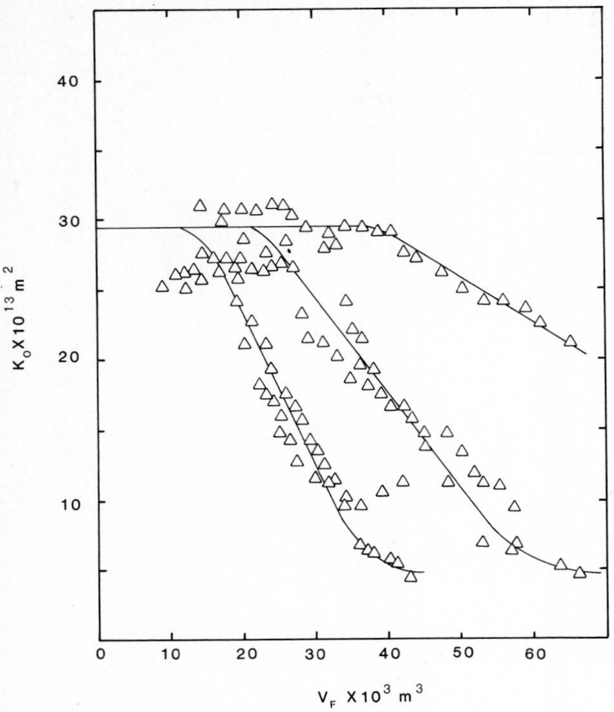

Fig. 19. Cake septum permeabilities throughout each filtration for all the filtration experiments using a Whatman No. 4 filter paper in the statistical analysis.

For the statistical analysis, a single value of the septum permeability must be used. This presents no problem for the constant value obtained with the non-clogging Whatman No. 3 filter paper but for the Whatman No. 4 filter paper an average value, defined by

$$\bar{K}_0 = \frac{1}{V_F}\int_0^{V_F} K_0 dV_F \tag{177}$$

is used. The data for the statistical analysis is shown in Table 4.

Analysis of Variance and Statistical Deductions

The average porosity data shown in Table 4 indicate, without benefit of statistical analysis, that the operating factors of media selection, pressure, and slurry concentration do not affect the average cake porosity. However, the internal pressure distributions shown in Fig. 16 and Fig. 17 indicate that there is a difference in cake structure. Consequently, the average porosity is a parameter that is insensitive not only to the operating factors but also to cake structure.

TABLE 4

Response of K_o and ε^*_α to the operating factors of slurry concentration, media selection and pressure.

MEDIUM: WHATMAN NO. 3				MEDIUM: WHATMAN NO. 4			
OPERATING FACTORS		RESPONSE		OPERATING FACTORS		RESPONSE	
s	$\dot{P}_c \times 10^{-3}$ Nm^{-2}	$K_o \times 10^{12}$ m^2	ε^*_α	s	$\dot{P}_c \times 10^{-3}$ Nm^{-2}	$\overline{K}_o \times 10^{12}$ m^2	ε^*_α
0.0517	34.5	2.75	0.378	0.0512	34.5	1.72[a]	0.375
0.0501	103.4	2.99	0.382	0.0508	103.4	1.98[a]	0.380
0.0995	103.4	3.15[a]	0.389	0.0998	103.4	1.26	0.380
0.0983	34.5	2.91[a]	0.392	0.0995	34.5	1.47	0.376
REPLICATES				REPLICATES			
0.0749	68.9	2.89	0.383	0.0739	68.9	1.32	0.385
0.0748	68.9	2.88	0.379	0.0745	68.9	1.75	0.385
0.0757	68.9	2.86	0.381	0.0739	68.9	1.20	0.381
0.0762	68.9	2.91	0.384	0.0745	68.9	1.39	0.378
				0.0752	68.9	1.79	0.383

[a]Average value for two experimental runs.

The analysis of variance for the cake septum permeability is shown in Table 5. The variance ratio column is obtained by divid-

TABLE 5

Analysis of variance for the cake septum permeability.

EFFECT	MEAN DIFFERENCE $\times 10^{13}$ m^2	VARIANCE $\times 10^{26}$ m^4	DEG. OF FREEDOM	VARIANCE RATIO	F-TEST[a] %
MAIN EFFECTS					
MEDIUM, M	13.4	360.	1	89.4	0.0031
PRESSURE, $\dot{P}_c$	1.33	3.53	1	0.88	38.1
SLURRY, s	- 1.63	5.33	1	1.32	28.8
INTERACTIONS					
$M \times \dot{P}_c$	1.12	2.52	1	0.63	45.5
Mxs	3.21	20.6	1	5.12	5.82
$\dot{P}_c \times s$	- 1.17	2.75	1	0.68	43.6
$M \times \dot{P}_c \times s$	1.18	2.78	1	0.69	43.4
ERROR[b]		4.03	7		

[a]The significance levels are from a program given by Volk (ref.27).
[b]Variances for s, $M\dot{P}_c$, $M\dot{P}_c xs$ are pooled with replicates to obtain the estimate of error variance.

ing each variance by the error variance. This variance ratio, together with the numerator and denominator degrees of freedom, are used in a program described by Volk (ref. 27) to obtain the F-test signifigance level or the probability that both variances in the variance ratio estimate the same population variance, (i.e., the null hypothesis that none of the operating factors affect the permeability response). For a Type I error of 2.5%, the only signifigant operating factor is media selection.

Initial Deposition Rate and Septum Permeability

The net effect of changing the operating factors of slurry concentration and pressure in the statistical analysis is to change the rate of deposition of cake particulates. The initial deposition rate of solids on the septum could affect the septum permeability. If Equation (165) is rearranged

$$L = \left(\frac{A}{\mu}\right)(K_0 J_0)\dot{P}_c \dot{V}_F^{-1} \tag{178}$$

and differentiated with respect to filtrate volume, then the expression for the deposition rate is

$$\frac{dL}{dV_F} = \frac{A}{\mu}\left[(K_0 J_0)\left(\dot{P}_c \frac{d\dot{V}_F^{-1}}{dV_F} + \dot{V}_F^{-1}\frac{d\dot{P}_c}{dV_F}\right) + \dot{P}_c \dot{V}_F^{-1}\frac{d(K_0 J_0)}{dV_F}\right] \tag{179}$$

The initial deposition rate

$$\frac{dL}{dt}\bigg|_{t=0} = \frac{dL}{dV_F}\dot{V}_F\bigg|_{V_F=0} = \frac{A}{\mu}(K_0)_0(J_0)_0\left[\frac{d\dot{P}_c}{dV_F}\right]\bigg|_{V_F=0} \tag{180}$$

is obtained by evaluating Equation (179) when the filtrate volume and cake pressure drop are zero.

Fig. 20 shows the initial value of the septum permability as a function of the initial deposition rate as calculated from Equation

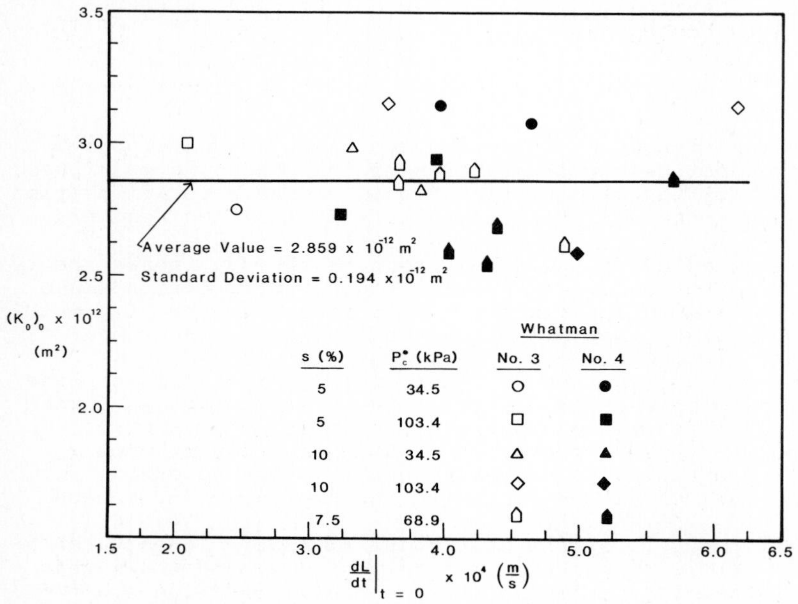

Fig. 20. Initial septum permeabilities, $(K_0)_0$, as a funtion of initial deposition rate for the Whatman No. 3 and No. 4 filter papers.

(180). The data shown in Fig. 20 corroborates the conclusion from the statistical analysis that the septum permeability is independent of the slurry concentration and cake pressure drop.

Implications of the Statistical Analysis

The conclusion from the statistical analysis that the average porosity is not affected by the slurry concentration and pressure provides an explanation for the initial curved portion of the cake length versus filtrate volume shown in Fig. 14 and Fig. 15.

At the start of the filtration, the slurry enters the filter chamber and is diluted by the clear filtrate that is initially present. During this initial period, the slurry concentation in the filter chamber increases from zero to the slurry concentration of the feed. Since the porosity of the cake is independent of the slurry concentration and pressure as deduced from the statistical analysis, the change in the slope, Equation (157), of cake length versus fitrate volume is due solely to the changing slurry concentration.

The extrapolation of the linear portion of the cake length versus filtrate volume intersects the abscissa at a point where the filtration would have begun if the filter chamber were filled initially with a slurry concentration equal to the feed concentration. The fact that the linear extrapolation of the reciprocal rate curve intersects the abscissa at the same point also supports this conclusion.

All of the previous data on cake length and reciprocal rate versus filtrate volume have been corrected to account for this initial change in slurry concentration and have been presented as filtrations that started with the filter chamber concentration equal to the feed concentration.

CONCLUSIONS

The averaging procedure provides a technique for developing equations that describe the interaction of the phases in multiphase systems. These multiphase equations provide a basis for filtration theory and replace the empirical approach that has been used previously.

The effort expended on the development of the multiphase theory is compensated for by a decrease in the experimental effort since the theory is developed for classes of slurries. Conversely, the empirical analysis does not distinguish between substances and hence, must be verified for each individual slurry.

Constitutive equations can be incorporated into the multiphase equations for each phase. This permits the development, as demonstrated here, of a filtration theory for soluble or elastic or non-deformable particulate phases in a Newtonian liquid. The procedure can be extended to arbitrary constitutive equations for either phase. This cannot be done with the empirical filtration analysis.

Dimensional analysis of the multiphase filtration theory for a non-deformable particulate phase in a Newtonian liquid shows that the dominant forces in the liquid phase momentum equation are gravity, pressure, and interfacial drag forces. The viscous forces are negligble which implies that in a one-dimensional filtration the velocity profile in the cake is flat.

For a one-dimensional filtration of a non-deformable particulate phase in a Newtonian liquid, the average porosity is constant and the local porosity is a function of only fractional cake position. The local porosity distribution determines the axial velocity profiles for both phases and the non-deformable particulate phase acts as simple conductor of momentum from the liquid phase to the septum.

In addition to the constitutive equations for both phases, which in this case specify a non-deformable particulate phase in a Newtonian liquid, two additional constitutive relations, or equivalent experimental information, must be available before predictive filter design or operation can be undertaken. The equivalent experimental information is the axial pressure and porosity profiles since these can be measured more easily than the permeability and velocity distributions.

The filtrate rate equation is obtained by evaluating the one-dimensional momentum equation at the exit of the filter cake where the particulate velocity is zero. The result indicates that the permeability at the cake septum interface is the controlling factor in determining the filtrate rate. This prediction is experimentally confirmed by comparing the results of filtrations in which a fine porosity non-clogging filter paper and a course porosity clogging filter paper are used.

A non-clogging septum will always give parabolic behavior but a clogging septum can give either parabolic or non-parabolic behavior. The only certain way to ascertain whether a septum is clogging is to measure the internal pressure profiles from which the septum permeability can be determined.

The initial curvature in the reciprocal rate versus filtrate volume is due to the rapid increase in the cake pressure drop. The intercept of the reciprocal rate data is simply the initial reciprocal rate through the clean septum.

Integration of the area under the pressure versus filtrate volume gives the energy dissipatd by the filtration process. For the data shown here, the non-clogging septum dissipates 46% more energy and 133% more power than the clogging septum.

Media selection affects filtrate clarity, time of filtration, energy and power requirments, and filter size. Judicious selection of a filter media is an important design and operational consideration.

The average porosity is independent of slurry concentration, pressure, and media selection. It is not a sensitive indicator of the internal structure of the filter cake.

The cake septum permeability is independent of slurry concentration and pressure, or equivalently, the initial cake deposition rate. This permeability is affected primarily by media selection.

ACKNOWLEDGEMENT

This work has been supported by the National Science Foundation, Grants ENG-7606224 and CPE-8007523, the Chemical Engineering Department, and the Office of the Coordinator of Research, University of Akron.

All of the following graduate students, Mr. Kenneth Gray, Mr. Ming Shen, Mr. Raymond Collins, Dr. Ismail Tosun, Mr. William Bridges, Mr. Steven Bybyk and Mr. Jagadeeshan Raviprakash have made signifigant contributions to this research.

NOMENCLATURE

a = interfacial area per unit volume, Equation (102)

A	=	filter area
A_D	=	surface area of averaging volume
A_α	=	area of A_D which coincides with the α-phase
A_β	=	area of A_D which coincides with the β-phase
$A_{\alpha\beta}$	=	area of the $\alpha\beta$-interfaces inside the averaging volume, V_D
$A_{\beta\alpha}$	=	area of the $\alpha\beta$-interfaces inside the averaging volume, V_D, $A_{\beta\alpha} = A_{\alpha\beta}$
D_{ij}	=	binary diffusivity for ij-system
f	=	an arbitrary function of position and time
$\underline{F}$	=	interphase drag force per unit volume, Equation (62)
Fr_α	=	α-phase Froude Number, Table 1
Fr_β	=	β-phase Froude Number, Table 1
g	=	magnitude of gravitational acceleration
$\underline{g}$	=	gravitational acceleration
$\underline{g}^*_\alpha$	=	$\bar{\underline{g}}^\alpha / g$
$\underline{g}^*_\beta$	=	$\bar{\underline{g}}^\beta / g$
G	=	shear modulus, Equation (72)
G	=	slope of cake length versus filtrate volume, Equations (156), (157)
$H(x-a)$	=	unit step function, Equation (16)
h	=	an arbitrary function of position and time
$\underline{I}$	=	identity tensor
$\underline{j}_i$	=	mass flux of species i relative to the mass average velocity
$\underline{j}_{i\alpha}$	=	the value of $\underline{j}_i$ in the α-phase
J_o	=	$(\partial \dot{P}^*_\alpha / \partial \xi)\|_{\xi=0}$, dimensionless pressure gradient at the septum
$(J_o)_o$	=	initial value of J_o
$\underline{k}$	=	rectangular cartesian unit vector
$\dot{k}_{loc}$	=	local mass transfer coefficient, Equation (101)
K	=	permeability, Equation (69)
K_o	=	permeability at the cake-septum interface, ξ=0, Equation (164)
$(K_o)_o$	=	initial value at t=0 of the cake-septum permeability at ξ=0
$\bar{K}_o$	=	mean value of K_o, Equation (177)
$(K_oJ_o)_o$	=	initial value of the (K_oJ_o) product

ℓ	=	dimensionless parameter in the pressure function, Equation (167)
L	=	cake length
$\dot{L}$	=	cake deposition rate or the speed of the cake slurry interface
m	=	dimensionless parameter in the pressure function, Equation (167)
M_α	=	mass of liquid α-phase in the filter cake
M_β	=	mass of particulate β-phase in the filter cake
n	=	parameter in the pressure function, Equations (167), (168)
$\underline{n}_\alpha$	=	unit normal vector directed out of the α-phase, Fig. 2
$\underline{n}_\beta$	=	unit normal vector directed out of the β-phase, Fig. 2
$\underline{n}_{\alpha\beta}$	=	unit normal vector at the $\alpha\beta$-interface directed out of the α-phase
$\underline{n}_{\beta\alpha}$	=	unit normal vector at the $\alpha\beta$-interface directed out of the β-phase
Nd_α	=	interphase drag force / α-phase viscous force, Table 1
Np_α	=	α-phase pressure force / α-phase viscous force, Table 1
Nr_β	=	β-phase inertial force / β-phase deformation force, Table 1
P	=	liquid phase thermodynamic pressure
P	=	applied filtration pressure, Equation (142)
$\dot{P}_\alpha$	=	local measured pressure, Equation (140)
P_0	=	pressure at the cake-septum interface
$\dot{P}_c$	=	measured pressure drop across the cake, Equation (142)
$\dot{P}_{cmax}$	=	maximum measured pressure drop across the cake
P_α^*	=	dimensionless pressure, $\bar{P}^\alpha/\dot{P}_c$, Equation (128)
$\dot{P}_\alpha^*$	=	dimensionless local measured pressure, $\dot{P}_\alpha/\dot{P}_c$
$\underline{r}$	=	position vector of a point in the averaging volume, Fig. 1
r_i	=	mass rate of production of species i by chemical reaction
$r_{i\alpha}$	=	α-phase mass rate of production of species i by chemical reaction
Re_α	=	α-phase Reynolds Number, Table 1
t	=	time
t^*	=	dimensionless ratio of time to the characteristic time
$\underline{T}_\omega$	=	tortuosity, Equation (97)
u_i	=	components of the strain vector, $\underline{u}$
$\underline{v}$	=	velocity vector

$\underline{v}_\alpha$ = local α-phase velocity

$\underline{v}_\beta$ = local β-phase velocity

$\underline{v}_1$ = local velocity of species 1

$\underline{v}_2$ = local velocity of species 2

$\bar{v}_z^\alpha$ = mass weighted average of the z-component of the α-phase velocity

$\bar{v}_z^\beta$ = mass weighted average of the z-component of the β-phase velocity

$\langle v \rangle_{\alpha z}$ = the z-component of the phase average α-phase velocity

$\langle v \rangle_{\beta z}$ = the z-component of the phase average β-phase velocity

$\langle v \rangle_{\alpha z,L}$ = $\langle v \rangle_{\alpha z}$ evaluated at the cake slurry interface

$\underline{v}_\alpha^*$ = dimensionless ratio of $\langle v \rangle_\alpha^\alpha$ to a characteristic velocity

$\underline{v}_\beta^*$ = dimensionless ratio of $\langle v \rangle_\beta^\beta$ to a characteristic velocity

V_D = averaging volume, Fig. 1

V_F = filtrate volume

$\dot{V}_F$ = filtrate rate

$\dot{V}_F^{-1}$ = reciprocal filtrate rate

V_α = volume of the α-phase in the averaging volume

V_β = volume of the β-phase in the averaging volume

$\underline{w}$ = velocity of the αβ-interface

W = work dissipated on the separation process of filtration

x = rectangular cartesian coordinate

x_i = rectangular cartesian coordinates, i = 1,2,3

$\underline{x}$ = position vector of the centroid of the averaging volume, Fig. 1

$\underline{x}_{\alpha\beta}$ = position vector of the αβ-interface

z = rectangular cartesian coordinate

Greek

$\underline{\gamma}$ = strain tensor

γ_{ij} = the ij component of the strain tensor

γ_α = distribution function, Equation (2)

$\delta(x-a)$ = Dirac delta function

ε_α = local volume fraction α-phase

ε_β = local volume fraction β-phase

$\varepsilon_{\alpha,L}$ = local volume fraction of the α-phase at the cake slurry interface

ε_α^* = average α-phase porosity, Equation (152)

ε_β^* = $(1 - \varepsilon_\alpha^*)$

λ = Lame constant

μ = α-phase viscosity

ξ = $z \,/\, L(t)$, fractional cake position

$\underline{\xi}$ = position vector in the ξ-coordinate system, Fig. 1

π_o = stress experienced by the septum

$\underline{\underline{\pi}}$ = stress tensor

$\underline{\underline{\pi}}_\beta$ = β-phase stress tensor

$\underline{\underline{\pi}}_\beta^*$ = dimensionless β-phase stress, $\underline{\underline{\pi}}_\beta \,/\, \pi_o$

$\bar{\pi}_{zz}^\beta$ = mass weighted average of the zz-component of $\underline{\underline{\pi}}_\beta$

$\dot{\pi}_{\beta zz}$ = zz-component of the β-phase stress tensor, Equation (141)

$\dot{\pi}_{\beta zz}^*$ = dimensionless zz-component of the β-phase stress tensor, $\dot{\pi}_{\beta zz} \,/\, \dot{P}_c$

ρ = density

ρ_i = density of species i

ρ_α = density of the fluid α-phase

ρ_β = density of the particulate β-phase

$\rho_{i\alpha}$ = density of species i in the α-phase

$\rho_{i\beta}$ = density of species i in the β-phase

$\underline{\underline{\tau}}$ = stress tensor

ϕ = dilatation

ψ = arbitrary scalar function

$\underline{\psi}$ = arbitrary vector function

ω_i = mass fraction of species i

Operators

$\langle\ \rangle_\alpha$ = phase average, Equation (7)

$\langle\ \rangle_\alpha^\alpha$ = intrinsic α-phase average, Equation (8)

$\bar{\ }^{\alpha}$ = mass weighted average, Equation (14)

$\hat{}\alpha$ = area average, Equation (66)

$\tilde{}\alpha$ = deviation, Equation (30)

$\underline{\nabla}_x$ = del-operator for the x-coordinate system, Equation (18)

$\underline{\nabla}^*$ = $L\underline{\nabla}$, dimensionless gradient operator

∇^{*2} = $L^2\nabla^2$, dimensionless Laplacian operator

REFERENCES

1 H.C. Weber and R.L. Hershey, Ind. Eng. Chem., 18 (1926) 341-344.
2 B.F. Ruth, G.H. Montillon, and R.E. Montonna, Ind. Eng. Chem., 25 (1933) 76-82.
3 M.S. Willis and I. Tosun, Chem. Eng. Sci., 35 (1980) 2427-2438.
4 M. Hassanizadeh and W.G. Gray, Adv. Water Resources, 2 (1979) 131-134.
5 M. Hassanizadeh and W.G. Gray, Adv. Water Resources, 2 (1979) 191-203.
6 W.G. Gray and P.C.Y. Lee, Int. J. Multiphase Flow, 3 (1977) 333-340.
7 M.J. Lighthill, Fourier Analysis and Generalized Functions, Cambridge Univ. Press, London, 1958.
8 P.A.M. Dirac, The Principles of Quantum Mechanics, Clarendon Press, Oxford, 1959.
9 R.B. Bird, W.E. Stewart and E.N. Lightfoot, Transport Phenomena, Wiley, New York, 1960.
10 J.C. Slattery, Momentum Energy and Mass Transfer in Continua, McGraw-Hill, New York, 1972, Sec. Ed. Krieger, Malabar, FL, 1981.
11 I. Tosun, Application of Multiphase Theory to Filtration, Ph.D. Dissertation, Univ. of Akron, Akron, OH (1977).
12 C. Truesdell and R.A. Toupin, Classical Field Theories, In S. Flugge (Ed), Handbuch der Physic, 3/1 Springer-Verlag, Berlin, 1960.
13 S. Whitaker, Ind. Eng. Chem., 61 (1969) 14-28.
14 L.A. Segel, SIAM Rev., 14 (1972) 547-571.
15 M. Shirato, M. Sambuichi and S. Okamura, AIChE J., 9 (1963) 599-603.
16 M. Shirato, T. Aragaki, K. Ichimura and N. Ootsuji, J. Chem. Eng. (Japan), 4 (1971) 172-177.
17 E. Hatschek, J. Soc. Chem. Ind., 27 (1908) 538-544.
18 B.F. Ruth, G.H. Montillon and R.E. Montanna, Ind. Eng. Chem., 25 (1933) 153-161.
19 A. Rushton and P.V.R. Griffiths, Filtr. and Sep., Jan/Feb (1972) 81-89.
20 H.P. Grace, AIChE J., 2 (1956) 307-336.
21 P.M. Heertjes, Chem. Eng. Sci., 6 (1959) 269-276.
22 P.M. Heertjes, Trans. Instn Chem. Engrs, 42 (1964) T226-T274.
23 E. Kehat, A. Lin and A. Kaplan, I&EC Proc. Des. & Dev., 6 (1967) 48-55.
24 J.G.A. Bitter, World Filtr. Congr., 1 (1979) 25-34.
25 T.D. Murphy, Chem. Eng., 84 (June 6,1977) 168-182.
26 G.E.P. Box, W.G. Hunter and I.S. Hunter, Statistics for Experimenters, Wiley, New York, 1978.
27 W. Volk, Engineering Statistics with a Programmable Calculator, McGraw-Hill, New York, 1982.

THEORETICAL ASPECTS OF HYDROCYCLONE FLOW

M.I.G.Bloor and D.B.Ingham
Department of Applied Mathematical Studies, University of Leeds,Leeds,England

CONTENTS

ABSTRACT

A theoretical analysis of various aspects of the fluid mechanics of the hydrocyclone is presented. The development of a much simplified model for the flow in the main body of the hydrocyclone is explained and examined critically, and the results compared with experimental data. The success of this simple rotational flow model in adequately predicting the fluid velocity distribution allows an analysis of the boundary layers to be carried out. Approximate methods based on a Pohlhausen approach are developed and in certain cases supported by very accurate full scale numerical solutions of the boundary layer equations. Having obtained a fairly full picture of the fluid mechanics of the system, a method for calculating a separating efficiency for small dense particles is given on the basis that the particles do not affect the fluid flow. However, the influence of the dense solid particles on the flow is then considered and both approximate and accurate numerical calculations are carried out in an attempt to gain some insight into the mechanisms of blockage at high concentration levels.

1. INTRODUCTION

1.1 General considerations

Over many years a great deal of effort has been devoted to obtaining a better working knowledge of hydrocyclones. This has been due in no small part to the apparent simplicity of operation and design inherent in this device. Conceptually the principle of operation is straightforward. Fluid containing particles of a different density, or even two fluids, is injected tangentially at high speed into a vessel of circular cross section. The high rotational velocities so generated produce centrifugal accelerations which cause the particles to move relative to the fluid and therefore offer the possibility of separation or classification subject to some means of collecting the distinct phases. The very high radial accelerations produced allow for a relatively rapid migration of particles thus allowing a large rate of volume flux through the equipment.

The very simplicity of this operation has probably dictated the course followed by many investigations. The fact that the mechanism of particle separation seemed straightforward together with the readily available information on particle drag at low concentration levels allowed intuition to play a significant part in the development of parameter groupings relevant to the operation of the hydrocyclone. This in turn was supported by extensive experimental investigations of the separation efficiency, the meaning of which, although it can be defined in a variety of ways, is adequately conveyed by its title. Not surprisingly, with this volume of data available and the intuitive understanding of the role played by various parameters it was possible to develop quite successful empirical formulae which predicted the performance of hydrocyclones. It is not the purpose of the present paper to give a review of this work and for information the reader is referred to Bradley (ref.1) where an exhaustive and detailed discussion of this important aspect of hydrocyclone research is given.

As the subject progressed and hydrocyclones came to be widely used, problems in operation arose and the need for a more fundamental study became apparent. It was recognised that a better understanding of the fluid mechanics was required and it was probably at this stage that the complexity of the flow problem became manifest. This, and again the fact that the hydrocyclone operated by centrifugal separation, dictated that early theoretical investigations concentrated on the azimuthal velocity distribution normally called the spin velocity . Fontein and Dijksman (ref.2), recognising that in the outer part of the hydrocyclone the spin velocity was likely to be that associated with a free vortex whilst near the axis solid body rotation existed, suggested a series relationship of the form $K_1/R + K_2R$ for the distribution, where R is the distance from the axis and K_1 and K_2 are constants. No values of K_1 and K_2 were given and clearly near the axis the wrong term dominates the expression. This type of relationship would certainly be of value if K_1 and K_2 were allowed to take different values in different regions of R . In particular, for R small K_1 must be taken to be zero.

On the experimental side, Kelsall (ref.3) carried out some careful measurements of the flow pattern in a conical hydrocyclone using an ingenious technique of measuring the trajectories of suspended alumina particles. This excellent scientific work supplied the type of information which was much needed to obtain some insight into the fundamentals of the fluid mechanics of the hydrocyclone. Indeed, using Kelsall's data for a velocity component, Rietema (ref.4) solved the azimuthal component of the Navier Stokes equations assuming a constant eddy viscosity. His results show a strong similarity with the experimentally determined spin velocity profiles, but no detailed comparisons were made.

The emphasis on the spin velocity, as has been mentioned, was understandable but with the benefit of hindsight it is probably true to say that this emphasis was largely misplaced. When the hydrocyclone is used to separate dispersed dense particles, later work (ref.5) has shown that it is the spin velocity in the outer region of the cyclone which is of significance.

Of course it is in this region where this velocity component is most easily determined theoretically because generally speaking viscous losses have not become significant. However, when a dispersed light phase is to be separated, the magnitude of the velocity near the axis is of paramount importance.

At this stage it is worth describing briefly the typical geometry of a conical hydrocyclone and discussing qualitatively the nature of the flow within it. With reference to Fig.1, it can be seen that the main body of the cyclone is a conical vessel, usually of small included angle, surmounted by a cylindrical section with a lid. A tube known as the vortex finder is located centrally through the lid and at the top of the cylindrical section are one or more inlet nozzles which direct the feed tangentially into the upper part of the body. At the apex of the cone there is an orifice through which some of the fluid entering the cyclone is allowed to leave. This underflow, as it is called, is taken to be a small fraction of the total flow in all that follows. Most of the fluid, the overflow, leaves through the vortex finder.

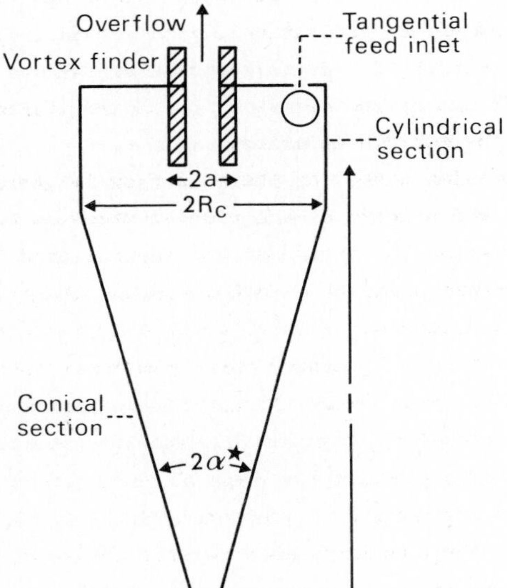

Fig.1 Typical hydrocyclone geometry

On entry to the hydrocyclone fluid passes through a nozzle to generate high speeds and then goes through what is effectively an expansion where streamlines diverge and quite significant losses may occur. Also the fluid has to adjust to the no slip conditions at the solid boundaries across boundary layers and, again, losses will arise. Account is usually taken of these losses by regarding the effective spin velocity at entry as a proportion (less than unity) of the actual entry spin velocity. The constant of proportionality is called the loss factor. The fluid then spirals down the outer part of the hydrocyclone and, since most must leave through the vortex finder, spirals up to the inner region around the axis. In the absence of any losses, so that angular momentum is conserved, the reduction in radius of curvature of the streamlines in the horizontal planes cause the spin velocity to increase and so increase the radial accelerations. The very high fluid speeds which arise near the axis cause low pressures and usually a gas liquid interface exists. If this core region communicates to the atmosphere then an air core is formed. Otherwise, if the pressure is sufficiently low, the liquid vaporises and a vapour core is formed. In the upper part of the hydrocyclone the rapid motion down the wall caused a toroidal vortex to be formed which usually extends beyond the lower end of the vortex finder. Clearly some radial motion must exist and this will play a significant part in determining whether a particle will be separated or not since the radial acceleration simply gives an indication of the migration of a particle relative to the surrounding fluid. The means of removal of the separated dense particles from the device relies on the intricate mechanics of the boundary layers on the solid boundaries.

Because the Reynolds number of the main flow is generally very large, boundary layers, in which inertial and viscous stresses balance, form on the walls. In the main flow the motion of the fluid around the axis is maintained by a pressure gradient directed towards the axis. Because the pressure is constant across a boundary layer, and within the boundary layer spinning motion is attenuated through the action of viscosity, an imbalance between pressure and inertial stresses due to spin arise. This imbalance causes fluid in the boundary layer to be driven in the direction of the pressure gradient giving rise to secondary flows. In the case of the boundary layer on the side wall, a component of the radial pressure gradient drives fluid through the boundary layer towards the vertex of the cone and thus assists in the removal of dense solid particles through the underflow. However, the boundary layer on the lid drives fluid which has just entered towards the vortex finder where it escapes to the overflow carrying with it particles which have not been subjected to the full separating mechanism of the hydrocyclone. This is commonly referred to as the 'short circuit flow' or 'leakage effect'.

No mention has yet been made of the influence of the particles on the flow. This is an aspect of hydrocyclone operation which has received little attention although certain practical guidelines based on experimental work do exist. It is nevertheless very important to consider this effect as the type of dischage at the underflow is a major practical consideration in certain applications. As the proportion of solid/liquid discharged at the underflow is increased the flow changes from a vortex discharge, where the mixture leaves in the form of a violent spray, to a 'rope' or 'sausage' discharge in the form of a rotating almost solid spiral. Eventually the hydrocyclone may become overloaded and the discharge is a near solid stream exhibiting little or no spiral motion.

Clearly, the influence of the particles on the flow will be most marked where the concentration is highest, that is in the side wall boundary layer, part or all of which is contained in the underflow. Some understanding of the influence that particles might have on the flow can be gained by recourse to elementary mechanics. The effect will be significant when the particle concentration and the flow is such that an interchange of momentum between the fluid and the particles can take place. This will arise where large accelerations exist and here again the sidewall boundary layer is a region to be considered. Concentrations in the main flow of the hydrocyclone are usually such that this interaction need not be considered. However, the influence of the particles on the spin velocity profile could be felt at quite low concentrations for the reasons just given. This effect would be similar to an increase in effective viscosity of the fluid.

The purpose of this article is to present an overall view of the theoretical approach to the hydrocyclone problem with particlar reference to the work of the authors and associates. In other words this is not to be regarded as an exhaustive review of the theoretical work in this area but rather a development of ideas leading to a more complete understanding of the intriguing fluid mechanics of the hydrocyclone. In the interests of trying to achieve a reasonably concise exposition only work which is directly relevant to this development will be discussed. Also, although sufficient analysis and arguments based on the authors' earlier papers will be presented to achieve a cohesive structure, the reader will necessarily be referred to the papers for some of the details.

1.2 Outline of problem

Having described in general terms the type of flow which exists in a hydrocyclone, the first move in any logical development must be to try to establish a model for the main flow. This then allows further developments in the more specialised areas to take place. Thus a solution for the main flow would be of great value if particularly simple functional forms for quantities such as the velocity components could be obtained. This necessarily means that the model will suffer from quite severe limitations but as long as these are recognised, progress can still be made.

The formulation of the main flow problem will, however, be presented fully for completeness and also to give a starting point for a complete full scale numerical solution. The simplification of the equations of motion, the boundary conditions and the elimination of the spin velocity to allow an analytic solution to be found will be explained. The rotational flow model previously developed by the authors (ref.6) will be presented and analysed critically, particularly with reference to the source of the vorticity which exists in the flow. The comparison with experimental data will be presented and the limitations of the theory discussed.

The simple model for the flow in the cross plane is then used to examine the spin velocity profile. A turbulent viscous model is necessary to produce transition from a free vortex flow near the side walls to solid body rotation near the axis. The functional forms derived and some ad hoc arguments about the form of the equations and boundary conditions, backed up by experimental evidence, are seen to imply a particular form for the eddy viscosity (ref.7). Solutions developed on this basis are shown to agree reasonably well with the experimental results. Futhermore, fluid dynamical arguments based on a Prandtl mixing length theory are presented to support the functional form of the eddy viscosity used. With this complete picture of the velocity components in the main flow it is then possible to analyse the boundary layers in the hydrocyclone.

The importance of both the side wall and top wall boundary layers in the operation of hydrocyclones has been described. Furthermore, the similarity of principal between the two flows has been explained and this is borne in mind in developing the boundary layer equations in a unifying form. It is argued that overall boundary layer properties such as skin friction and induced volume flux are required and consequently the momentum integral equations are obtained. The use of the Pohlhausen method in solving this type of equation is examined and the rules deduced to remove a good deal of the arbitrariness inherent in this method (ref.8). The full boundary layer equations are solved numerically wherever possible to very high accuracy and the results used to lend confidence to the guidelines laid down for the use of the Pohlhausen method.

The problem of the sidewall boundary layer is then solved (ref.9) using the approximate method, and the way in which the radial pressure gradient assists in driving fluid through this boundary layer is demonstrated. The effect of this on the required flux through the underflow is also mentioned. The boundary layer on the top wall is complicated by the presence of the toroidal vortex. A model is developed which takes account of this motion by reference to the flow just beneath the vortex which is accurately predicted by the main flow theory. Again, a Pohlhausen method allows the flux of fluid through this boundary to be evaluated and therefore gives an estimate of the leakage (ref.10). The influence of certain design parameters, e.g. the radius of the vortex finder, on this quantity is discussed. This completes the picture of the fluid mechanics within the limitations of the approximate theories put forward and attention can be directed towards a consideration of the particle motions.

The way in which individual small particles move in the flow is examined on the basis that the motions of the particles do not affect each other or the flow. Having established acceptable approximations for the motion, the efficiency of the hydrocyclone can be calculated and the effect of design parameters on this efficiency discussed. The assumption that particles do not interact is examined critically and unless the discharge at the underflow is of very low solids concentration it is deduced that the influence of the particles on the side wall boundary layer should be considered. One approach is to try to assess the overall effect of the particles by having a concentration dependent coefficient of viscosity. An examination of Laverack's side wall boundary layer analysis (ref.11) on this basis reveals that while some useful practical information can be deduced from his results, a clearer understanding of the detailed interaction of particles with the flow is required.

The full equations of motion, taking account of particles, are presented and the method of solution following Laverack (ref.12) is discussed. The detailed picture of the particle transport within the boundary layer presented by this solution is used to discuss the development of an approximate method based on Pohlhausen's technique. It is hoped that this model will give guidance on the use of a hydrocyclone with high solid concentration levels in the underflow and go some way to understanding the mechanism of blockage.

2. BASIC FLOW MODEL

2.1 General equations

A theoretical investigation of the fluid dynamics of a hydrocyclone necessarily involves some treatment of the Navier Stokes equations. For the moment, it will be assumed that the effect of the solid particles on the flow is insignificant, although this point will be considered later in this article. Furthermore, the fluid will be assumed to behave in a Newtonian manner and have a constant coefficient of viscosity when turbulence is absent. When the effects of turbulence are taken into account, by means of an eddy viscosity, the spatial variation of the viscosity must be allowed for. For constant viscosity, the Navier Stokes equation for steady incompressible flow becomes

$$\underline{\nabla}\,(p/\rho + \frac{1}{2}q^2 + \chi) - \underline{q}\wedge\underline{\omega} = -\nu \text{ curl } \underline{\omega} \tag{2.1}$$

where $\underline{q}$ is the fluid velocity, $\underline{\omega}$ = curl $\underline{q}$ is the vorticity and p, ρ, χ and ν represent pressure, density, potential energy per unit mass and kinematic viscosity respectively. In practice χ would simply be the potential energy per unit mass due to gravity and can be eliminated by regarding p as the pressure measured in excess of hydrostatic pressure. In addition, conservation of mass gives

$$\underline{\nabla} \cdot \underline{q} = 0 . \tag{2.2}$$

A typical geometry for a conical hydrocyclone is shown in Fig.1 and to obtain a complete solution to the problem it would be necessary to solve eqns. (2.1) and (2.2) subject to no slip conditions on $\underline{q}$ at the solid boundaries together with inlet conditions at the nozzle and exit conditions at the underflow and overflow. Clearly the problem as it stands is intractable and it is necessary to make some simplifying assumptions before any progress can be made.

Although it is impossible to arrange for an axially symmetric tangential inlet flow and consequently the flow near the inlet is fully three dimensional, the geometry of the system is such that the assumption of axial-symmetry in the flow over most of the interior of the hydrocyclone seems reasonable. Also the Reynolds number of the flow is generally high which means that the effects of viscosity are confined to boundary layers on the solid boundaries and possibly free shear layers in the flow.

If the flow may be regarded as inviscid it is necessary to distinguish between irrotational ($\underline{\omega} = 0$) and rotational flow. Firstly, if the flow is irrotational, there exists a velocity potential φ defined, to within an arbitrary additive constant, by

$$\underline{q} = \underline{\nabla}\,\varphi . \tag{2.3}$$

Hence by virtue of eqn. (2.2)

$$\nabla^2 \varphi = 0 , \tag{2.4}$$

At a stationary boundary with normal in the direction $\underline{n}$, the normal component of velocity is zero, i.e.

$$\underline{\nabla}\,\varphi \cdot \underline{n} = 0 . \tag{2.5}$$

Hence, for this case, eqn.(2.4) must be solved subject to the boundary condition, eqn.(2.5).

Secondly, if the vorticity in the flow is not identically zero, taking the curl of eqn.(2.1) with $\nu = 0$ yields

$$\text{curl } (\underline{q} \wedge \underline{\omega}) = 0 \quad . \tag{2.6}$$

This equation takes a particularly simple form for either plane two dimensional flow or axially symmetric flow without swirl. In each of these cases, the vorticity is perpendicular to the plane of the motion and the simplification arises because the flow variables do not change in this direction. The way in which this is used will be discussed more fully later.

It will be convenient to introduce more than one coordinate system for a study of the basic flow. Spherical polar coordinates (r,θ,λ) are used with the origin at the vertex of the cone, the axis being $\theta = 0$ and the surface of the cone $\theta = \alpha^*$. The azimuthal coordinate is λ . Also, using the same origin, cylindrical polar coordinates (R,λ,z) can be defined where $R = r \sin\theta$ and the z-axis coincides with $\theta = 0$. Fig.2 illustrates the coordinate systems.

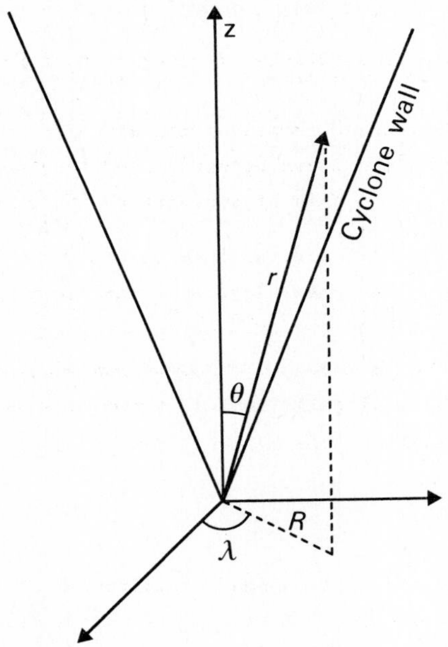

Fig.2 The coordinate systems.

On the basis that the important region of flow in the hydrocyclone is the conical section , which is also the region where the assumption of axial symmetry is likely to be valid, attention is now confined to a purely conical vessel where allowance must be made for the practical aspects of fluid entry.

2.2 Flow in the (r,θ) plane

Without reference to any specific vorticity generating mechanism in the fluid at entry to or within the hydrocyclone it would seem reasonable to look for a solution in which vorticity is absent. Using eqn.(2.3) the velocity potential φ can be defined by

$$q_r = \frac{\partial\varphi}{\partial r} \quad \text{and} \quad q_\theta = \frac{1}{r}\frac{\partial\varphi}{\partial\theta} \tag{2.7}$$

where the suffices denote the particular components of $\underline{q}$.

The continuity equation (2.4) then becomes

$$\frac{\partial}{\partial r}\left(r^2 \frac{\partial\varphi}{\partial r}\right) + \frac{1}{\sin\theta}\frac{\partial}{\partial\theta}\left(\sin\theta \frac{\partial\varphi}{\partial\theta}\right) = 0$$

which has a solution (ref.13)

$$\varphi = [CP_n(\cos\theta) + DQ_n(\cos\theta)] \cdot \left[r^n + \frac{E}{r^{n+1}}\right]$$

where P_n and Q_n are Legendre functions of order n; C, D and E are arbitrary constants.

Ignoring, for the moment, the presence of the air core around the axis of the cyclone, the boundary conditions can be determined from eqn.(2.5). At the surface of the cone and on the axis of symmetry the velocity component, q_θ, is zero, so that

$$\frac{\partial\varphi}{\partial\theta} = 0 \text{ at } \theta = 0 \text{ and } \theta = \alpha^* .$$

In addition, there can be no terms in the solution involving $Q_n(\cos\theta)$ and $r^{-(n+1)}$ because of their singular nature on the axis of symmetry and at $r = 0$ respectively. Hence, since the equation for φ is linear

$$\varphi = \sum_n C_n P_n(\cos\theta)\, r^n$$

where $P'_n(\cos\alpha^*) = 0$ to satisfy the boundary condition. The summation is over all n for which this is true. In all cases of practical importance, α^* is small and the lowest admissible value of n is then quite large. The resulting variation in the velolcity with r unfortunately bears no relationship at all to experimental findings and consequently this type of solution must be abandoned. The only assumption which seems likely to have been illfounded is that of irrotationality. Hence, it is necessary to look for a rotational solution.

The simplification in the equations of motion shown in eqn. (2.6) may be used to good effect. Using eqn. (2.2), eqn. (2.6) reduces to

$$\left(q_r \frac{\partial}{\partial r} + \frac{q_\theta}{r} \frac{\partial}{\partial \theta}\right)\left(\frac{\omega_\lambda}{r\sin\theta}\right) = 0 \tag{2.8}$$

where

$$\omega_\lambda = \frac{1}{r}\frac{\partial}{\partial r}(rq_\theta) - \frac{1}{r}\frac{\partial q_r}{\partial \theta}$$

is the λ component of vorticity. Eqn.(2.8) simply states that in an axisymmetrical flow, the ratio $\omega_\lambda/(r\sin\theta)$ is constant along each streamline since the intensity of each vortex ring changes during convection in proportion to its radius.

A stream function ψ is introduced to satisfy the continuity equation so that:

$$\frac{\partial\psi}{\partial r} = -r\sin\theta\, q_\theta \qquad \text{and} \qquad \frac{\partial\psi}{\partial\theta} = r^2\sin\theta\, q_r$$

Eqn.(2.8) can then be integrated to give

$$-\frac{1}{r^2\sin\theta}\left[\frac{1}{\sin\theta}\frac{\partial^2\psi}{\partial r^2} + \frac{\partial}{\partial\theta}\left(\frac{1}{r^2\sin\theta}\frac{\partial\psi}{\partial\theta}\right)\right] = f(\psi)\,, \tag{2.9}$$

since the operator $q_r\partial/\partial r + (q_\theta/r)(\partial/\partial\theta)$ represents differentiation along a streamline.

Since the semi-angle of the cone α^* is small, the rates of change with respect to θ are much larger than those with respect to r so that eqn.(2.9) can be approximated by

$$\frac{\partial}{\partial\theta}\left(\frac{1}{\theta}\frac{\partial\psi}{\partial\theta}\right) = -r^4\theta\, f(\psi) \tag{2.10}$$

for small θ, taking $\sin\theta \simeq \theta$.

The function $f(\psi)$ represents the distribution of vorticity in the fluid. The source of this vorticity cannot be specified at this stage but it would seem likely that the geometry of the hydrocyclone itself ensures that a certain type of distribution must exist. In view of this any assumptions about the functional form of f must be of a fairly general nature. Also it seems reasonable to attempt a solution of the problem using a Pohlhausen type of approach. That is to say, that plausible assumptions about the functional forms of the solution must be made so that, for example, boundary conditions are satisfied. Then the equation of motion is used to determine parameters left unspecified in these functional forms.

Assuming that the leading term in $f(\psi)$ behaves like $F\psi^\delta$, where F

and δ are constants, an approximate solution can be sought in the form

$$\psi = Br^{\beta}\theta^{\gamma}(\alpha^* - \theta) .$$

The constants F, B, β, γ and δ are yet to be determined from the equation of motion and the boundary conditions.

The term in $(\alpha^*-\theta)$ is linear so that the velocity at the wall is tangential, finite and non-zero; in practice this is the velocity just outside the boundary layer at the wall. Providing $\gamma > 1$ the boundary condition for q_θ on the axis is also satisfied. The expression for ψ is really a truncated series in θ . Higher order terms could be retained but the solution obtained from the leading terms shows surprisingly good agreement with experiment.

Substituting the assumed form for ψ in eqn.(2.10) and retaining just the first two terms in the expansion of $(\alpha^*-\theta)^{\delta}$, the constants β, γ, and δ can be determined. In fact

$$\beta = \gamma = \frac{3}{2}$$

$$\delta = -\frac{5}{3}$$

$$B^{8/3} = \frac{4}{3} F \alpha^{*-8/3}$$

so that

$$\psi = B(r\theta)^{3/2}(\alpha^*-\theta) \tag{2.11}$$

The velocity components q_r and q_θ are then determined in the form

$$\left.\begin{aligned} q_r &= \frac{1}{2} B(r\theta)^{-1/2}(3\alpha^*-5\theta) \\ &\text{and} \\ q_\theta &= -\frac{3}{2} B r^{-1/2} \theta^{1/2} (\alpha^*-\theta) \end{aligned}\right\} ; \tag{2.12}$$

For comparison with experimental work, it will be found more convenient to use cylindrical polar coordinates in which ψ can be written.

$$\psi = BR^{3/2} (\alpha^* - R/z) \tag{2.13}$$

to within the order of approximation used previously. The velocity components in the R and z directions are then given by

$$q_z = \frac{1}{R}\frac{\partial\psi}{\partial R} = \frac{1}{2} B R^{-1/2}(3\alpha^* - 5R/z) \tag{2.14}$$

which of course is the same as q_r because θ is small, and

$$q_R = -\frac{1}{R}\frac{\partial\psi}{\partial z} = -\frac{BR^{3/2}}{z^2} \tag{2.15}$$

It is worth noticing that at a given value of z for small values of R the radial component of velocity q_R varies like $R^{3/2}$ whereas the axial velocity q_z varies like $R^{-1/2}$. This means that the flow is approximately axial for small values of R so that the solution would still be a reasonable approximation even with an air core present. This also eliminates the physically unrealistic presence of an infinite velocity on the axis itself.

When the hydrocyclone is used to separate a dispersed dense phase, the amount of fluid removed at the underflow represents only a small fraction of the total flow through the device. This, together with the leakage from the boundary layer on the vortex finder to the overflow, may safely be ignored at this stage. The volume flux Q through the hydrocyclone can then be used to determine the constant B in the solution. It has already been mentioned that the validity of the solution cannot extend to the upper part of the hydrocyclone but should certainly represent a good approximation to the flow near the end of the vortex finder. Here, as has already been observed for small R, the flow is almost entirely axial and so would approximately satisfy the necessary boundary condition on a very thin walled tube coaxial with the hydrocyclone. Of course, in practice the tube is thick walled and it is necessary to examine how a radius for an equivalent thin walled tube might be deduced. It is only possible to argue intuitively that the presence of the wall would cause the approaching fluid to move towards the air core rather than form part of the recirculating flow. On this basis, the outer radius of the vortex finder would seem to be the appropriate length to choose.

Hence, using eqn.(2.13) at the level of the end of the vortex finder, a distance ℓ from the vertex of the cone say, B is determined by

$$2\pi B a^{3/2}(\alpha^* - a/\ell) = Q . \tag{2.16}$$

In order to compare the theoretical results with the corresponding experimental findings of Kelsall (ref.3) the parameters in the system are chosen to be those appropriate to Kelsall's apparatus.

Using eqn.(2.13) the streamline pattern in the (r,θ) plane can be obtained and is shown in Fig.3.

The horizontal and vertical components of velocity are shown in Figs. 4a and 4b respectively at various levels in the hydrocyclone. These were obtained from eqns.(2.14) and (2.15). Included in the figures are the experimental results of Kelsall for comparison. It is seen that the theoretical results overestimate the horizontal velocities; this is due to the neglect in the theory of the leakage near the vortex finder and the absence of an air core.

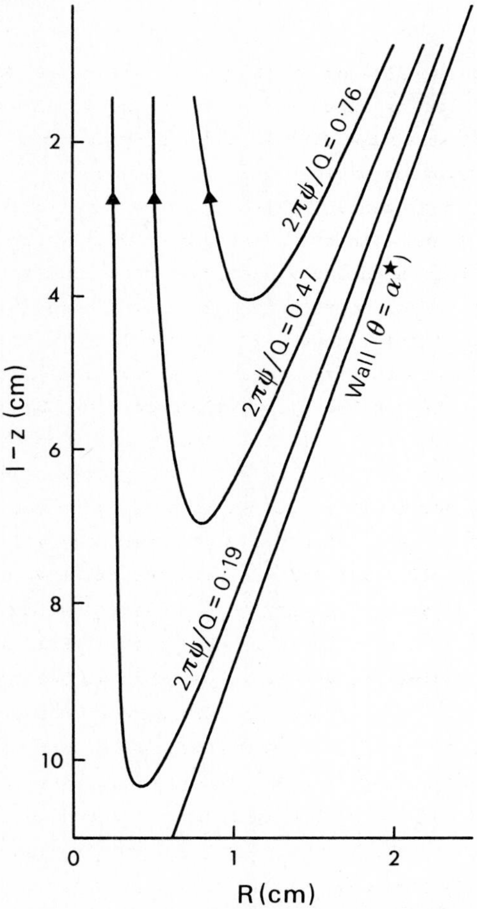

Fig.3 Streamline pattern

It is clear that the general behaviour of the flow is accurately predicted by the theory. In particular the locus of zero axial velocity has been found experimentally to be $\theta = 0.61\alpha^*$ (see ref.1), whereas the theory predicts $\theta = 0.60\alpha^*$ as can be seen from eqn.(2.14). Kelsall found this line to be $\theta = 0.57\alpha^*$ but attributed this lower value of the factor to the presence of an extra long vortex finder.

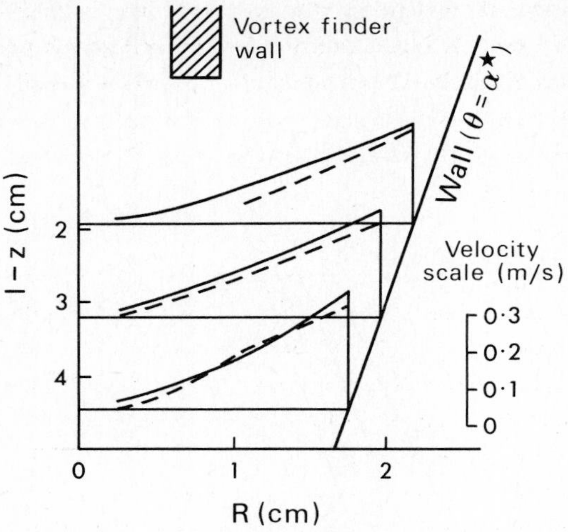

Fig.4a The horizontal velocity component at various levels in the hydrocyclone

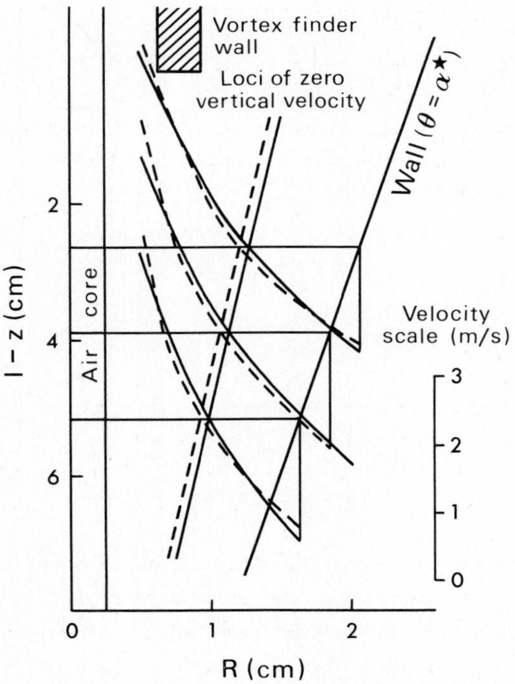

Fig.4b The vertical component of velocity at various levels in the hydrocyclone

It is probably instructive at this stage to return to the important question of the source of the vorticity represented by the function $f(\psi)$ in eqn.(2.9). Using the fact that $F = 3/4(B\alpha^*)^{8/3}$ and eqn.(2.16) it can be seen that

$$f(\psi) = \frac{3}{4}\left(\frac{Q}{2\pi}\right)^{8/3} \psi^{-5/3} a^{-4} .$$

The unboundedness of $f(\psi)$ is neither as unrealistic nor as impractical as might at first be supposed. On the wall of the hydrocyclone where $\psi = 0$ the neglect of viscosity means that a slip velocity is permissible. In the real problem the vorticity is distributed through the boundary layer but in this idealised case the boundary layer is of zero thickness giving rise to what is effectively a vortex sheet on the wall.

The lower bound on $f(\psi)$ is worth considering as a means of establishing the importance of the vorticity which might be generated at the inlet. When $\psi = Q/2\pi$, the minumum value of $f(\psi)$ is $3Q/(2\pi a^4)$. At the inlet, components of vorticity in all coordinate directions are almost certainly generated. However, for the flow to be axially symmetric, it is obvious that the only non-zero component of vorticity is in the azimuthal direction. Indeed, it is likely that the loss factors generally associated with the spin velocity in a hydrocyclone are associated with the dissipation of the component of vorticity in the cross plane. To estimate the order of magnitude of the surviving vorticity, consider an inlet channel with typical linear dimension, D, then the vorticity in the cross plane of the inlet is $O(Q/D^3)$ and the component of this in the azimuthal direction is $O(Q/D^{5/2}R_c^{1/2})$ where R_c is the radius of the hydrocyclone and is assumed to be large compared with L so that $(D/R_c)^4 << 1$. Remembering that $f(\psi)$ represents the vorticity divided by the radius of the vortex ring, the contribution to f from the inlet vorticity is $O(Q/D^{5/2}R_c^{3/2})$. Thus the ratio of the contribution from the inlet vorticity to the total vorticity is of $O(a^4/D^{5/2}R_c^{3/2})$. For Kelsall's experimental data this ratio is about 20% and therefore indicates that the inlet vorticity can have a significant effect on the overall fluid flow in a hydrocyclone and indeed could be used to improve the separating efficiency. However, it is certainly not clear that all of the vorticity is generated at the inlet and further investigation of the full Navier Stokes equation is required.

2.3 The spin velocity

In the previous section, a solution for the flow in the cross plane was found and no reference was made to the spin velocity q_λ. Of course, with the assumptions set out in that section together with the inlet conditions, a solution for q_λ was implicit in the scheme. This can be seen by returning to eqn.(2.1) with $\nu = 0$. The λ component of this equation in cylindrical polar coordinates is

$$q_R \frac{\partial q_\lambda}{\partial R} + q_z \frac{\partial q_\lambda}{\partial z} + \frac{q_R q_\lambda}{R} = 0 , \tag{2.17}$$

and the general solution is

$$Rq_\lambda = g(\psi) , \tag{2.18}$$

where g is an arbitrary function of ψ . In other words the moment of momentum of a fluid element is unchanged as it moves through the flow. This obviously arises because viscosity, which can destroy the momentum of the fluid element, has been ignored. For a hydrocyclone all the fluid elements enter with approximately the same moment of momentum so that $g(\psi)$ is equal to a constant, say A . Then

$$q_\lambda = A/R \tag{2.19}$$

and the spin velocity distribution is the familiar free vortex. In practice, of course, the infinite velocity on the axis $R = 0$ is not achieved because viscosity takes a hand and in fact for R sufficiently small solid body rotation exists, which is clearly a solution of the full Navier Stokes eqn.(2.1) because the shear is zero.

In the outer region of the cyclone, just outside the boundary layer on the wall, where viscosity is unimportant, the spin velocity is reasonably accurately predicted by eqn.(2.19) providing account is taken of the substantial losses already referred to in the previous section. When the hydrocyclone is used to separate a dispersed dense phase it can be shown that the spin in the outer region is of paramount importance and eqn.(2.19) is adequate. The spin near the axis is largely irrelevant since a dense particle in this region would be lost to the overflow. This will be discussed more fully in a later section. On the other hand, if the migration of light dispersed particles is of interest, the very high spin achieved near the centre of the cyclone is extremely important and account must then be taken of the effect of viscosity on the spin velocity profile.

Again assuming axially symmetric flow, the equation which determines the spin velocity q_λ is obtained from the equivalent of eqn.(2.1) where allowance is made for the possible spatial variation of the viscosity. In cylindrical polar coordinates this is

$$\rho \frac{q_r}{R} \frac{\partial}{\partial R} (q_\lambda R) + \rho q_z \frac{\partial}{\partial z} q_\lambda = (\frac{\partial}{\partial R} + \frac{2}{R}) \left[\mu (\frac{\partial q_\lambda}{\partial R} - \frac{q_\lambda}{R}) \right] + \frac{\partial}{\partial z} (\mu \frac{\partial q_\lambda}{\partial z}) \tag{2.20}$$

If q_R and q_z can be regarded as known, this equation must be solved subject to boundary conditions on q_λ . In the outer region of the cyclone, q_λ is given by eqn.(2.19) and ignoring for the moment the presence of the air core

$$q_\lambda = 0 \quad \text{at} \quad R = 0 \tag{2.21}$$

In the previous section, it was demonstrated that viscosity should not have any appreciable effect on the flow in the cross plane. It is therefore worthwhile to consider how a spin velocity distribution different from that given by eqn.(2.19) might influence this cross plane flow. The inviscid momentum equations in the R and z directions are respectively

$$q_R \frac{\partial q_R}{\partial R} + q_z \frac{\partial q_R}{\partial z} - \frac{q_\lambda^2}{R} = - \frac{1}{\rho} \frac{\partial p}{\partial R} \tag{2.22}$$

and

$$q_R \frac{\partial q_z}{\partial R} + q_z \frac{\partial q_z}{\partial z} = - \frac{1}{\rho} \frac{\partial p}{\partial z} \tag{2.23}$$

It can be seen that if q_λ is a function of R only, then a modified pressure p^* defined by

$$p^* = p - \int^R \frac{q_\lambda^2}{R} dR \tag{2.24}$$

may be introduced so that eqns.(2.22) and (2.23) become

$$q_R \frac{\partial q_R}{\partial R} + q_z \frac{\partial q_R}{\partial z} = - \frac{1}{\rho} \frac{\partial p^*}{\partial R} \tag{2.25}$$

and

$$q_R \frac{\partial q_z}{\partial R} + q_z \frac{\partial q_R}{\partial z} = - \frac{1}{\rho} \frac{\partial p^*}{\partial z} \tag{2.26}$$

respectively. These equations could be regarded as representing a flow in which the spin velocity is absent and therefore corresponds to the problem solved in section (2.2) with $A \equiv 0$.

Returning now to the solution of equation (2.20), it should be noticed that the boundary conditions (2.19) and (2.21) are independent of z and it is therefore consistent to assume that q_λ is a function of R only. The experimental evidence gives very strong confirmation of this assumption. This is extremely convenient because it means that, in view of eqns. (2.24),(2.25) and (2.26), the velocity comonents q_R and q_z given by eqns.(2.14) and (2.15) are consistent with eqn.(2.20).

Thus eqn.(2.19) reduces to

$$- \rho B \frac{R^{1/2}}{z^2} \frac{d}{dR} (q_\lambda R) = \frac{\partial}{\partial R} \left[\mu \left(\frac{dq_\lambda}{dR} - \frac{q_\lambda}{R}\right) \right] + \frac{2\mu}{R} \left(\frac{dq_\lambda}{dR} - \frac{q_\lambda}{R}\right) \qquad (2.27)$$

where eqns.(2.14) and (2.15) have been used for the velocity components q_R and q_z. It can be seen from eqn.(2.27) that the variation in μ must be of the form

$$\mu = \frac{f(R)}{z^2} \qquad (2.28)$$

A physical justification of this form will be developed later. Using eqn.(2.28) for μ, the partial differential eqn.(2.31) reduces to the ordinary differential equation

$$-\rho B R^{1/2} \frac{d}{dR} (q_\lambda R) = \frac{d}{dR} \left[f(R) \left(\frac{dq_\lambda}{dR} - \frac{q_\lambda}{R}\right) \right] + \frac{2f(R)}{R} \left(\frac{dq_\lambda}{dR} - \frac{q_\lambda}{R}\right) \qquad (2.29)$$

Since q_λ is independent of z, it will be convenient to solve eqn.(2.29) at the level corresponding to $R = R_c$. Then the boundary condition (2.19) can be replaced by

$$q_\lambda = V_i \quad \text{at} \quad R = R_c$$

where V_i is the inlet spin speed suitably adjusted for losses. Although the boundary condition (2.21) ignored the presence of the air core, the experimental evidence indicates that the air core usually falls within the region of solid body rotation where the shear stress is zero. Consequently the only boundary condition which needs to be satisfied at the surface of the air core is that of continuity of velocity. Hence the presence of the air core is irrelevant to the spin velocity profile and the boundary condition (2.21) may be used with confidence.

If non-dimensional variables R' and q'_λ given by

$$R = R_0 R' \quad , \quad q_\lambda = V_0 q'_\lambda \qquad (2.30)$$

where R_0 is the radius of the hydrocyclone at the level $z = \ell$ and V_0 is the spin velocity at this level, eqn.(2.29) becomes

$$-LR'^{1/2}\frac{d}{dR'}(q'_\lambda R') = \frac{d}{dR'}\left[g(R')\left(\frac{dq'_\lambda}{dR'} - \frac{q'_\lambda}{R'}\right)\right] + \frac{2g(R')}{R'}\left(\frac{dq'_\lambda}{dR'} - \frac{q'_\lambda}{R'}\right) \tag{2.31}$$

The non-dimensional parameter L is given by

$$L = \frac{\rho B R_0^{5/2}}{\mu_t} \tag{2.32}$$

where

$$f(R) = \mu_t g(R') \tag{2.33}$$

and $\ell^2\mu_t$ is a typical viscosity, for example the mean value in the hydrocyclone. The parameter L is the Reynolds number based on the radial velocity and the radius of the hydrocyclone.

Thus eqn.(2.31) has to be solved subject to the boundary conditions

$$\left.\begin{array}{llll} q'_\lambda = 0 & \text{at} & R' = 0 \\ q'_\lambda = 1 & \text{at} & R' = 1 \end{array}\right\} \tag{2.34}$$

It is possible to solve eqn.(2.31) analytically for suitably chosen $g(R')$ and the advantage of this type of solution in possible developments involving this theory justifies such a choice.

Taking $g(R') = 1$, eqn.(2.31) reduces to

$$-LR'^{1/2}\frac{d}{dR'}(q'_\lambda R') = \frac{d}{dR'}\left[\frac{1}{R'}\frac{d}{dR'}(q'_\lambda R')\right]. \tag{2.35}$$

Eqn.(2.35) can be integrated twice to give

$$q'_\lambda = \frac{1}{R'}\frac{\gamma\left(\frac{4}{5},\frac{2}{5}LR'^{5/2}\right)}{\gamma\left(\frac{4}{5},\frac{2}{5}L\right)} \tag{2.36}$$

satisfying the boundary conditions (2.34). Here

$$\gamma(n,x) = \int_0^x t^{n-1}e^{-t}\,dt$$

is the Incomplete Gamma function.

It is worth noting that for sufficiently large L eqn.(2.36) can be written

$$q'_{\lambda} = \frac{1}{R'} \frac{\gamma(\frac{4}{5},\frac{2}{5} LR'^{5/2})}{\Gamma(\frac{4}{5})} \tag{2.37}$$

The solution for q'_{λ} given by eqn.(2.36) is plotted in Fig.5 for various values of the parameter L . Also shown for comparison are the experimental results of Kelsall, using series I operating conditions at a level 3 cm below the vortex finder. It can be seen that reasonable agreement is obtained when L is of the order of 100 and this implies that the eddy viscosity at this level is about 0.4 poise.

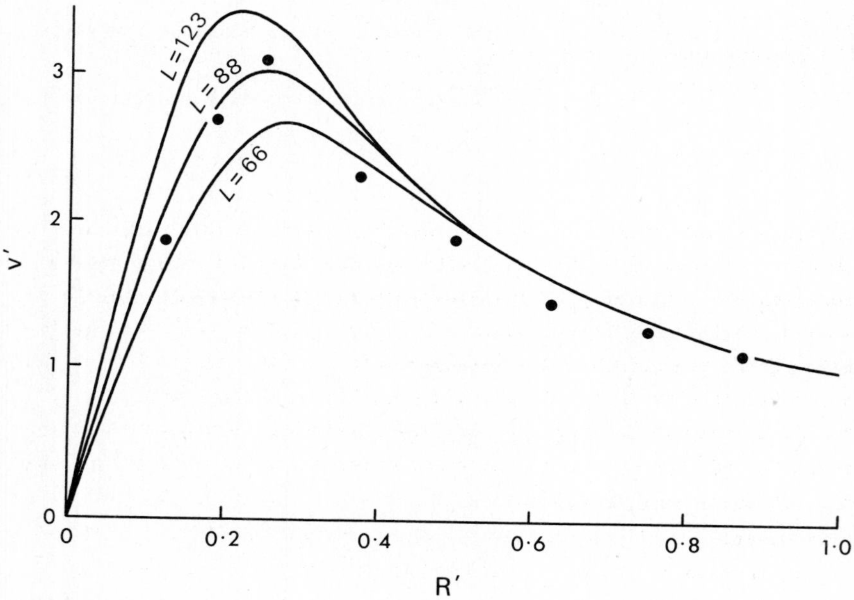

Fig.5 The velocity distribution across the hydrocyclone for various values of the parameter L.

An attempt is now made to develop a physically realistic model for the eddy viscosity which will give some insight into the turbulent transfer processes and possibly explain the required variation with z .

The basis of the present approach is the Prandtl mixing length model (see ref.14) which states that the eddy viscosity is given by

$$\mu = \rho \Lambda^2 \text{ x (rate of strain)}$$

and is applicable to unidirectional flows; Λ is the mixing length, in other words the length scale of the turbulent eddies. The analogous model for the flow of the present problem is

$$\mu = \rho\Lambda^2 \left|\frac{\partial q_\lambda}{\partial R} - \frac{q_\lambda}{R}\right| \tag{2.38}$$

This model would result in zero eddy viscosity in the region where there is no shear, that is near the centre of the hydrocyclone where solid body rotation exists. This seems unreasonable as turbulence does not vanish at points of zero rates of strain and it would result in solid body flow existing right across the hydrocyclone. Hence the model is modified by the addition of a constant term K which augments the shear so that

$$\mu = \rho\Lambda^2 \left|\frac{\partial q_\lambda}{\partial R} - \frac{q_\lambda}{R} + K\right| \tag{2.39}$$

It is necessary to determine the basis on which a realistic length scale, Λ, can be chosen. It has been suggested (see ref.15) that the length scale associated with the eddies in a turbulent flow is that associated with the boundary layer thickness. In the case of the hydrocyclone the boundary layer on the wall can be regarded as originating near the entry. This type of flow has been investigated by Rott and Lewellen (ref.16). Using a momentum integral method, they found that the boundary layer thickness on the wall is proportional to $z^{1/2}(\ell-z)^{1/2}$, and thus increases as z decreases to a point about half way down the hydrocyclone. It has already been noted that the analysis is greatly simplified if the viscosity varies like ℓ/z^2. This requires that the length scale Λ vary as ℓ/z, since in this case q_λ would be a function of R only (see eqn.(2.27)). Although these variations are not strictly comparable the main feature is in agreement, namely decreasing z increases Λ in the upper half of the hydrocyclone. In this region of interest these variations can be made to agree to within about 20%. It therefore seems reasonable to take a ℓ/z variation in Λ, bearing in mind the enormous simplifications so produced. Hence the form for the eddy viscosity is

$$\mu = \frac{M}{z^2} \left|\frac{dq_\lambda}{dR} - \frac{q_\lambda}{R} + K\right| \tag{2.40}$$

where M is a constant.

With reference to eqn.(2.28) and (2.33), eqn.(2.40) gives

$$g(R') = \left|\frac{dq'_\lambda}{dR'} - \frac{q'_\lambda}{R'} + S\right| \tag{2.41}$$

where

$$\mu_t = MV_0 / R_0$$

and

$$S = R_0 K / V_0 \tag{2.42}$$

Here S is a parameter which governs the size of the eddy viscosity in the region near the axis of the cyclone.

Substituting for $g(R')$ in eqn.(2.31) gives a non-linear second order ordinary differential equation for q'_λ which has to be solved subject to the boundary conditions (2.34).

Since this is a non-linear, two-point, boundary-value problem, the differential equation was solved indirectly by scaling the dependent and independent variables appropriately. The scaled equation was solved numerically using the Runge Kutta Merson method and the parameters S and L were deduced from the solution.

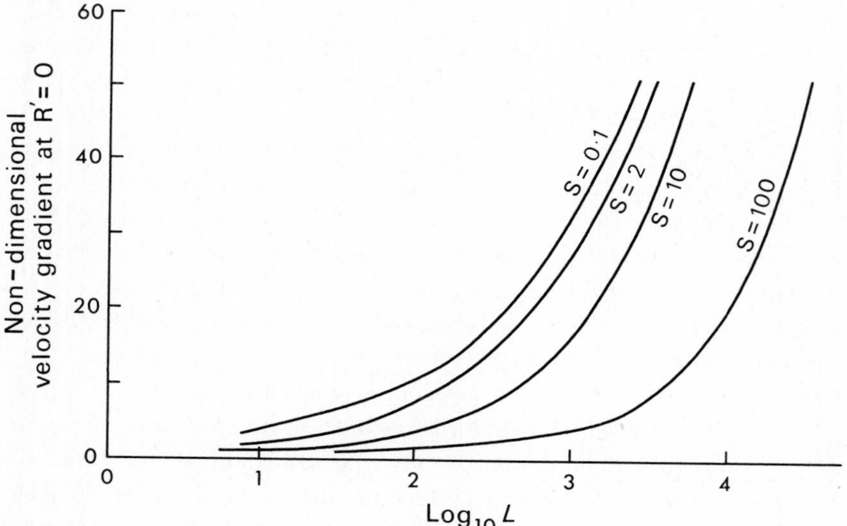

Fig.6 The variation of the non-dimensional velocity gradient at $R'=0$ plotted against $\log_{10} L$ for various values of S.

Fig.6 shows the variation of the non-dimensional velocity gradient at $R' = 0$ plotted against $\log_{10} L$ for various values of S. Thus given a physical problem where L and S are known the initial velocity gradient can be determined and eqn.(2.31) can be integrated directly to give the velocity distribution.

The experimental results of Kelsall form a basis for comparison. In order to obtain the relevant parameters for his system, use can be made of Fig.6. From Kelsall's results it can be seen that the non-dimensional velocity gradient at the centre of the hydrocyclone is approximately 16. Since in the present model the air core has been ignored, the appropriate value would be slightly higher. Consequently the curves are plotted for slopes ranging from 15 to 25. For the case of $S = 0.1$ this leads to values of L of the order of a few hundreds and these profiles are shown in Fig.7. The agreement with Kelsall's experimental results is again very reasonable. When $S = 10$ the range of L is from about 800 to 2000 and these results are shown in Fig.8. The agreement with Kelsall's experimental results is very slightly better near the maximum velocity but perhaps not quite as good elsewhere.

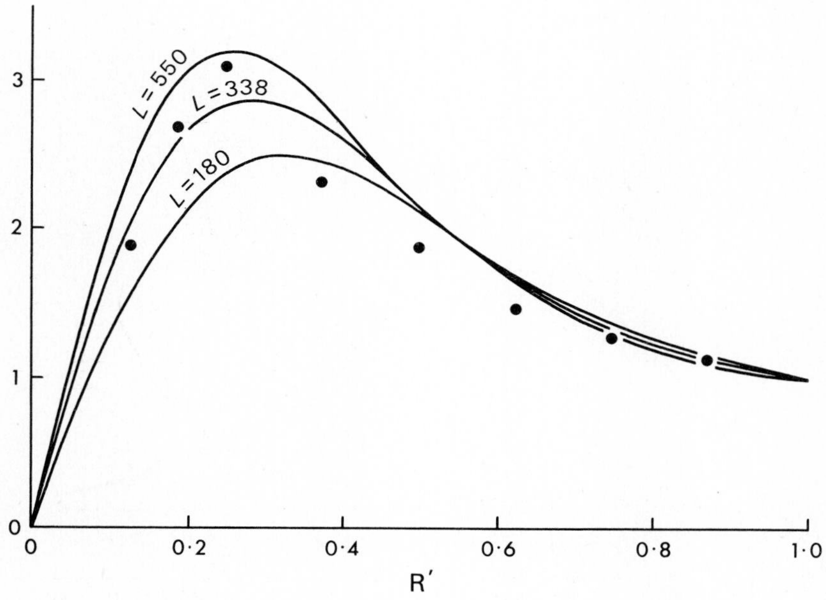

Fig.7 The velocity distribution across the hydrocyclone for various values of L with S=0.1

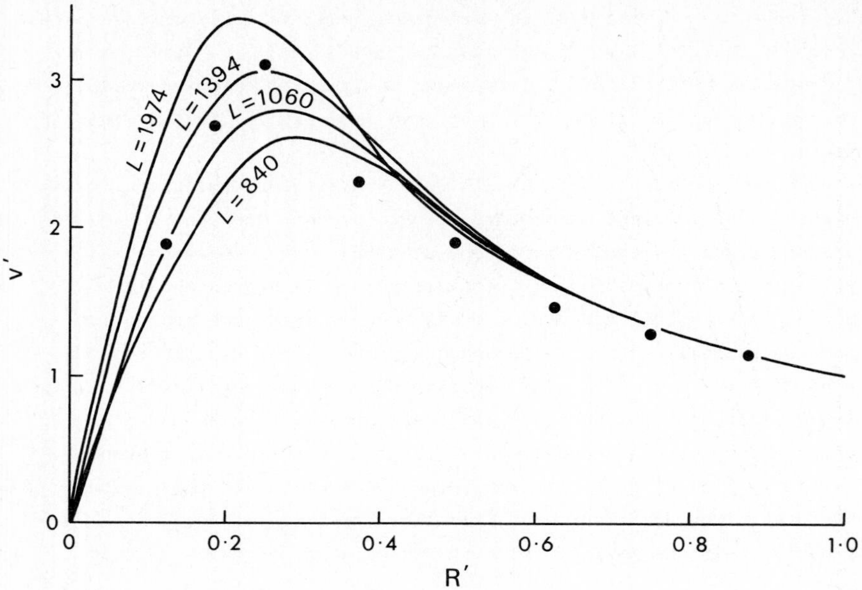

Fig.8 The velocity distribution across the hydrocyclone for various values of L with S=10.

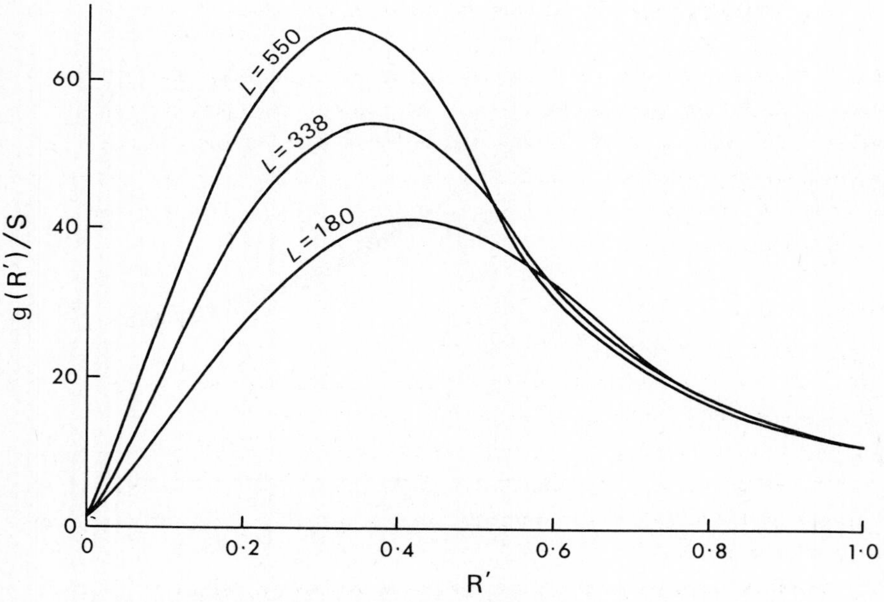

Fig.9 The distribution of the eddy viscosity across the hydrocyclone for various values of L with S=0.1.

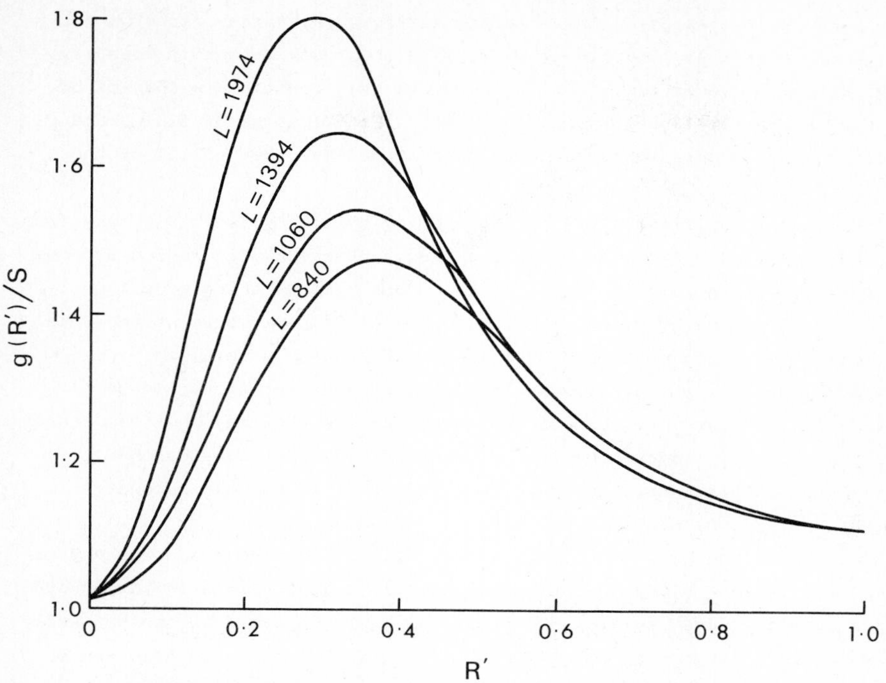

Fig.10 The distribution of the eddy viscosity across the hydrocyclone for various values of L with S=10.

In Figs. 9 and 10 the distribution of eddy viscosity across the hydrocyclone is shown for parameters S and L appropriate to Figs. 7 and 8 respectively. The maximum value of the viscosity in both of these cases, for Kelsall's data, is about 0.5 poise and this value is attained in the vicinity of the maximum spin velocity. For comparison, the simple solution given by eqn.(2.36) gives a similar value.

3. BOUNDARY LAYER FLOWS

3.1 General equations

Because the Reynolds number of the main flow is generally very large boundary layers are formed on the walls of the hydrocyclone. The spin velocity in the main flow of the hydrocyclone introduces a radial pressure gradient which, as previously explained, due to the imbalance in the wall boundary layers of the pressure and inertial forces produces secondary motions within the boundary layers.

In the boundary layer on the lid of the hydrocyclone (hereafter referred to as the top wall of the hydrocyclone) the pressure gradient serves to drive the fluid radially inwards. This fluid moves down the vortex finder and thence to the overflow. Particles which find their way into this boundary layer "leak" to the overflow. It is essential that the fluid mechanics of this boundary layer be investigated in detail as the leakage of fluid, and hence particles, by this short circuit route is of the order of 10% of the total flux.

In section 2 the flow in the conical region of the hydrocyclone away from the side wall of the hydrocyclone was analysed. Between this region and the top boundary layer there exist eddy flows. This recirculating eddy flow is driven by the fluid being drawn from the region of high pressure at the inlet to the low pressure region at the overflow. At present a detailed mathematical analysis of the flow in these eddies does not appear to be possible. This, therefore, complicates any analysis that may be attempted on the top wall boundary layer because the inviscid flow just outside the boundary layer is unknown. However, this eddy flow is of only secondary importance and will be discussed further in section 3.2.

In a similar manner to the top wall boundary layer flow the component of the pressure gradient along the hydrocyclone wall drives the fluid through the side wall boundary layer towards the underflow. In contrast to the top wall boundary layer, where the external radial motion is of secondary importance, the external radial motion in the side wall boundary layer effectively determines the order of magnitude of the thickness of the boundary layer. Although the radial component of velocity is smaller than the spin velocity at the edge of the side wall boundary layer it is of the same order of magnitude (see ref.3). Hence it is essential that this component of velocity be retained in the analysis. Although this complicates the problem somewhat, any solution derived ignoring this effect would be unacceptable in a fundamental study of the problem. It is essential that the flow in the side wall boundary layer is understood thoroughly since it is through this layer that the separated dense particles find their way to the underflow. All attempts at estimating the efficiency of hydrocyclones have been based, either directly or indirectly, on the assumption that if a solid particle reaches the boundary layer it will be eventually discharged through the underflow.

The essential features of the two boundary layers can be found by examining the boundary layer on the inside of a stationary cone, of semi angle σ^*, produced by a swirling flow around the axis of the cone and a flow parallel to the generators of the cone directed towards the vertex. The side wall boundary layer solution then being deduced from $\sigma^* = \alpha^*$ and the top wall boundary layer from $\sigma^* = \pi/2$.

A similar problem was studied by Taylor (ref.17) who considered the secondary motion through the boundary layer in a swirl atomiser due to a free vortex on the axis of the cone. A Pohlhausen technique was adopted in which the circumferential and meridional momentum integral equations determined the thickness of the boundary layer and the radial velocity peak in the boundary layer. Binnie and Harris (ref.18) took account of the radial flow in Taylor's problem, but they did not allow for the overshoot velocities in the boundary layer. Cooke (ref.19) extended Taylor's work by assuming the existence of two different boundary layer thicknesses for the radial and tangential velocities. However, the results due to this refinement were not as good as might be expected and hence the present work is restricted to only one boundary layer thickness. Rott and Lewellen (ref.16) have applied the momentum integral method to both laminar and turbulent boundary layer flows on bodies of various shapes. Outside the boundary layer, however, only a swirling motion was considered.

A similar problem to those described above is that of a laminar boundary layer on a fixed finite disc, induced by some external symmetrical flow. Burgraff, Stewartson and Belcher (ref.20) studied the boundary layer flow when the external flow was that due to a potential vortex, the axis of which was perpendicular to the disc. Starting from the edge of the disc they solved the boundary layer equations numerically using central finite differences and deduced that a double-structured boundary layer existed near the centre.

Because of the exact nature of the central finite difference method there is no doubt that this method is more accurate than the Pohlhausen, the accuracy of which depends to some extent on the profiles chosen. However, in some problems, it is impossible to solved the governing partial differential equations using central finite difference methods, e.g. for the flow on the top wall of the hydrocyclone where the external radial flow is unknown. In these circumstances there is no alternative but to employ an approximate method and an extension of the Pohlhausen method is probably the best approach. Further, in many boundary layer problems the detailed knowledge of the flow is not required but rather overall or average quantities are sufficient. In these circumstances the Pohlhausen method provides a very efficient and easy way of achieving this. Although much work has been done on the use of the Pohlhausen method with unidirectional external flow, very little rigorous analysis has been done when two components of velocity outside the boundary layer are involved. Hence it is important to use this method with caution and verification by difference methods is desirable. This ensures that guidelines used in the choice of velocity profiles have some firm foundation and thus, where no direct check is available, the results may be viewed with some confidence.

A coordinate system is set up in which s is measured along a generator of the cone from the vertex and n is measured normal to the surface of the cone and directed towards the axis of symmetry. The velocity components outside the boundary layer in the s and azimuthal directions are denoted by U and V respectively. Within the boundary layer, the equations of motion reduce to the usual boundary layer equations, which in this coordinate system are

$$u \frac{\partial u}{\partial s} + w \frac{\partial u}{\partial n} - \frac{v^2}{s} = -\frac{1}{\rho}\frac{\partial p}{\partial s} + \nu \frac{\partial^2 u}{\partial n^2} \qquad (3.1)$$

$$\frac{u}{s}\frac{\partial}{\partial s}(sv) + w \frac{\partial v}{\partial n} = \nu \frac{\partial^2 v}{\partial n^2} \qquad (3.2)$$

$$\frac{v^2}{s} \cot\sigma^* = -\frac{1}{\rho}\frac{\partial p}{\partial n} \qquad (3.3)$$

where u, v and w are the velocity componetns in the s, λ and n directions respectively.

In addition the continuity equation can be written

$$\frac{1}{s}\frac{\partial}{\partial s}(su) + \frac{\partial w}{\partial n} = 0 \qquad (3.4)$$

The boundary conditions appropriate to this situation are

$$u = v = w = 0 \qquad \text{at} \quad n = 0 \qquad (3.5a)$$

and at the outer edge of the boundary layer

$$u = U(s) \; , \quad v = V(s) \; . \qquad (3.5b)$$

Since the boundary layer is assumed to be thin, eqn.(3.3) shows that the pressure is effectively constant across this layer provided $s \sin\sigma^*$ is large compared with the boundary layer thickness. Hence eqn.(3.1) becomes

$$u \frac{\partial u}{\partial s} + w \frac{\partial u}{\partial n} - \frac{v^2}{s} = U\frac{dU}{ds} - \frac{V^2}{s} + \nu \frac{\partial^2 u}{\partial n^2} \qquad (3.6)$$

using the fact that just outside the boundary layer viscous effects are unimportant and the pressure gradient is given by the equation

$$U \frac{dU}{ds} - \frac{V^2}{s} = -\frac{1}{\rho}\frac{\partial p}{\partial s} \qquad (3.7)$$

Eqns.(3.3) and (3.6) can now be integrated across the boundary layer to obtain the momentum integral equations in the form

$$\frac{\partial}{\partial s}\left\{\int_0^\delta s\ u\ (u-U)dn\right\} + \frac{dU}{ds}\left\{\int_0^\delta s(u-U)dn\right\} + \int_0^\delta (V^2-v^2)dn = -\ \nu s\ \frac{\partial u}{\partial n}\bigg|_{n=0} \tag{3.8}$$

and

$$\frac{1}{s^2}\frac{d}{ds}\left\{\int_0^\delta s^2u\ (v-V)dn\right\} = -\ \nu\ \frac{\partial v}{\partial n}\bigg|_{n=0} \tag{3.9}$$

where $\delta = \delta(s)$ denotes the boundary layer thickness. In deriving eqns.(3.8) amd (3.9) use has been made of the fact that the shearing stress is zero at the edge of the boundary layer.

By integrating the continuity equation across the boundary layer one obtains

$$w\bigg|_{n=\delta} = -\ \frac{1}{s}\frac{\partial}{\partial s}\left\{\int_0^\delta su\ dn\right\} + U\frac{d\delta}{ds}$$

Eqns.(3.2),(3.4) and (3.6) subject to the boundary conditions (3.5) may now be solved using finite differences. Alternatively, eqns.(3.8) and (3.9) subject to the boundary conditions (3.5) may be solved using the Pohlhausen method.

3.2 Pohlhausen method

To use a Pohlhausen type of approach to solve the momentum integral equations it is necessary to choose the velocity profiles across the boundary layer. Generally speaking the more conditions satisfied by these profiles at the edge of the boundary layer and at the solid boundary, the more accurate are the results. However, before making some plausible assumptions about the profiles it is worthwhile considering the physical situation in some detail.

The effect of the swirling motion is to produce a pressure gradient, a component of which drives fluid through the boundary layer in a radial direction towards the vertex. This means that the velocity component, u, must take account of this effect as well as the effect of the external flow directed towards the vertex. The azimuthal component of velocity, v, simply increases from zero at the wall to its mainstream value. Hence the velocity components are expressed in the form

$$\left.\begin{aligned} \frac{u}{U} &= F(\eta,s) + \lambda_1(s)\, f_1(\eta) \\ &\text{and} \\ \frac{v}{V} &= g(\eta) \end{aligned}\right\} \quad (3.10)$$

where $\eta = n/\delta(s)$ is a similarity variable and the function $f_1(\eta)$ is introduced in order to take into account the induced flow through the boundary layer. The relative importance of this term will be determined by the function $\lambda_1(s)$, found from the solution to the moment integral equations.

The conditions which will be satisfied by the assumed profiles are, at n = 0

$$\left.\begin{aligned} \nu \frac{\partial^2 u}{\partial n^2} &= \frac{V^2}{s} - U\frac{dU}{ds} \;, \quad u = 0 \\ &\text{and} \\ \frac{\partial^2 v}{\partial n^2} &= 0 \;, \quad v = 0 \;. \end{aligned}\right\} \quad (3.11a)$$

Also at $n = \delta$

$$\left.\begin{aligned} \frac{\partial u}{\partial n} &= \frac{\partial^2 u}{\partial n^2} = \ldots = \frac{\partial^m u}{\partial n^m} = 0 \;, \quad u = U \\ &\text{and} \\ \frac{\partial v}{\partial n} &= \frac{\partial^2 v}{\partial n^2} = \ldots = \frac{\partial^m v}{\partial n^m} = 0 \;, \quad v = V \end{aligned}\right\} \quad (3.11b)$$

where m is an integer. Usually m has been taken to be large, see Wilkes (ref.21), Bloor & Ingham (ref.8), but recent work by Gaunt (ref.22) questions this assumption.

In view of the conditions (3.11) it is convenient to write $F(\eta,s)$ in eqn.(3.10) as

$$F(\eta,s) = f_0(\eta) + \lambda_2(s)\, f_2(\eta) \quad (3.12)$$

where $\lambda_2(s)$ will be adjusted so that the second of conditions (3.11a) is satisfied while f_0 accounts for the skin friction in the radial direction.

The momentum integral eqns. (3.8) and (3.9) can now be written

$$\frac{d}{ds}\left\{ s\,\delta\,U^2 \int_0^1 (f_0 + \lambda_1 f_1 + \lambda_2 f_2)(f_0 + \lambda_1 f_1 + \lambda_2 f_2 - 1)\, d\eta \right\}$$

$$+ U\frac{dU}{ds}\, s\,\delta \int_0^1 (f_0 + \lambda_1 f_1 + \lambda_2 f_2 - 1)\, d\eta + \delta\, V^2 \int_0^1 (1 - g^2)\, d\eta$$

$$= - \frac{\nu\, s\, U}{\delta}\,(f_0'(0) + \lambda_1 f_1'(0)) \tag{3.13}$$

and

$$\frac{1}{s^2}\frac{d}{ds}\left\{ s^2\, U\, V\, \delta \int_0^1 (f_0 + \lambda_1 f_1 + \lambda_2 f_2)(g - 1)\, d\eta \right\} = - \frac{\nu\, V}{\delta}\, g'(0) \tag{3.14}$$

Furthermore

$$\frac{\nu U}{\delta^2}\, f_2''(0)\,\lambda_2 = \frac{V^2}{s} - U\frac{dU}{ds} \tag{3.15}$$

providing $f''_0(0) = f''_1(0) = 0$.
To summarize, the conditions to be satisfied by the functions g, f_0, f_1 and f_2 are

$$\left.\begin{array}{l}
g\,(0) = 0,\ g'\,(0) = 1,\quad g''\,(0) = 0,\ g\,(1) = 1. \\[2ex]
\left.\dfrac{d^i g}{d\eta^i}\right|_{\eta=1} = 0, \quad i=1,\ldots,m; \\[2ex]
f_0(0) = 0,\ f'_0(0) = 1,\quad f''_0(0) = 0,\ f_0(1) = 1. \\[2ex]
\left.\dfrac{d^i f_0}{d\eta^i}\right|_{\eta=1} = 0, \quad i=1,\ldots,m; \\[2ex]
f_1(0) = 0,\ f'_1(0) = 1,\quad f''_1(0) = 0. \\[2ex]
\left.\dfrac{d^i f_1}{d\eta^i}\right|_{\eta=1} = 0, \quad i=0,\ldots,m; \\[2ex]
f_2(0) = f_2'(0) = 0,\quad f_2(0) = -2. \\[2ex]
\left.\dfrac{d^i f_2}{d\eta^i}\right|_{\eta=1} = 0, \quad i=0,\ldots,m;
\end{array}\right\} \tag{3.16}$$

The two ordinary differential eqns. (3.13) and (3.14) together with the algebraic eqn. (3.15) determine, subject to suitable boundary conditions, the variables δ, λ_1 and λ_2 with the f's and g prescribed. It is reasonable to assume that the boundary layer thickness is initially zero, i.e.

$$\delta = 0 \quad \text{at} \quad s = s_0 \tag{3.17}$$

where the suffix zero refers to the conditions at the start of the boundary layer.

Also referring to the eqns. (3.10), the flow through the boundary layer induced by the spin should initially be zero. Hence

$$\lambda_1 = 0 \quad \text{at} \quad s = s_0 \tag{3.18}$$

An examination of eqns. (3.13), (3.14) and (3.15) as $s \to s_0$, using conditions (3.17) and (3.18) reveals that the relationship

$$(I_{00} - I_0)\, g'(0) = (J_0 - I_0)\, f'_0(0) \tag{3.19}$$

must hold for consistency (see ref.8). Here the quantities I_{ij}, I_i and J_i are defined by the integrals

$$\left.\begin{aligned} I_{ij} &= \int_0^1 f_i\, f_j\, d\eta\,, \\ I_i &= \int_0^1 f_i\, d\eta\,, \\ J_i &= \int_0^1 f_i\, g\, d\eta\,, \end{aligned}\right\} \tag{3.20}$$

where the integers i,j may be 0, 1 or 2.

Now the values of $f'_0(0)$ and $g'(0)$ effectively determine the ratio of the components of skin friction in the s and λ directions near $s = s_0$. However, $f'_0(0)$ and $g'(0)$ can be chosen arbitrarily, the only constraint linking f_0 and g being eqn. (3.19). In order to resolve the arbitrariness and put the choice of the ratio $f'_0(0)/g'(0)$ on some sound basis it is necessary to look at the eqns. (3.1), (3.2) and (3.3) near $s = s_0$. This was done by Bloor and Ingham (ref.8) and it was found that both the azimuthal and radial components of velocity start near $s = s_0$ with Blasius profiles. Some details of this are shown in Section (3.3). This result is in genereal agreement with the findings of Sears (ref.23) in relation to the boundary layer on a yawed wing.

Of particular importance in the present context is that the ratio of the initial radial and azimuthal components of the skin friction is determined. Hence

$$f_0'(0) = g'(0) \tag{3.21}$$

Following the above guidelines it is possible to choose the forms of the assumed profiles. Hence when m=7

$$\left.\begin{aligned} f_0 &= g = 1 - (1-\eta)^8[1 + 7\eta + 28\eta^2 + 84\eta^3] \\ f_1 &= (1-\eta)^8\eta\,[1 + 8\eta + 36\eta^2] \\ f_2 &= -(1-\eta)^8\,\eta^2\,[1 + 8\eta\,] \end{aligned}\right\} \tag{3.22}$$

and when m=3

$$\left.\begin{aligned} f_0 &= g = 1 - (1-\eta)^4[1 + 3\eta + 6\eta^2 + 10\eta^3] \\ f_1 &= (1-\eta)^4\eta\,[1 + 4\eta + 10\eta^2] \\ f_2 &= -(1-\eta)^4\,\eta^2\,[1 + 4\eta\,] \end{aligned}\right\} \tag{3.23}$$

Although not strictly necessary, it seems reasonable on physical grounds to choose $g = f_0$, conditions (3.19) and (3.21) are then clearly satisfied. With the profiles and the forms U and V chosen eqns. (3.13) and (3.14) can be integrated numerically using condition (3.15) for λ_2 to yield f and λ_1. The importance of the consistency eqn. (3.19) is clearly shown in that a relaxation of this condition leads to wholly unrealistic results (see ref.8).

3.3 Finite-difference equations

It will be convenient to non-dimensionlise the flow variables according to the relations

$$s' = s / s_0$$

$$n' = n\,(-U_0 / \nu\, s_0)^{1/2}$$

$$u' = u / U_0$$

$$v' = v / V_0$$

$$w' = w\,(-s_0 / \nu\, U_0)^{1/2}$$

where the prime denotes a non-dimensional quantity.

Since the integration is started at $s = s_0$, $s' = 1$, and continued towards $s' = 0$ then the behaviour near $s' = 1$ has to be determined as for the Pohlhausen method. It is well known that the appropriate similarity variable near $s' = 1$ is

$$\zeta = n'/(1 - s')^{1/2} \tag{3.24}$$

In order to satisfy the continuity equation the non-dimensional stream function ψ' is introduced and is defined by

$$\frac{\partial \psi'}{\partial s'} = - s' w' \; ; \qquad \frac{\partial \psi'}{\partial n'} = s' u' \tag{3.25}$$

and writing

$$\left.\begin{aligned} \psi' &= (1 - s')^{1/2} H(s',\zeta) \\ s' v' &= G(s',\zeta) \end{aligned}\right\} \tag{3.26}$$

then eqns. (3.24) and (3.25) become

$$\frac{\partial H}{\partial \zeta}\left[s' \frac{\partial^2 H}{\partial \zeta \partial s'} - \frac{\partial H}{\partial \zeta}\right](1 - s')$$

$$+ \frac{\partial^2 H}{\partial \zeta^2}\left[\frac{1}{2} s'H - s'(1 - s') \frac{\partial H}{\partial s'}\right]$$

$$= s' - 1 - K^2(1 - s')(1 - G^2) + s'^2 \frac{\partial^3 H}{\partial \zeta^3} \tag{3.27}$$

and

$$\frac{\partial H}{\partial \zeta} \frac{\partial G}{\partial s'}(1 - s')s' + \frac{\partial G}{\partial \zeta}\left[\frac{1}{2} s'H - s'(1 - s') \frac{\partial H}{\partial s'}\right] = s'^2 \frac{\partial^2 G}{\partial \zeta^2} \tag{3.28}$$

where $K = -V_0/U_0$.

As $s' \to 1$, eqns. (3.27) and (3.28) reduce to

$$\frac{\partial^3 H}{\partial \zeta^3} + \frac{1}{2} H \frac{\partial^2 H}{\partial \zeta^2} = 0 \tag{3.29}$$

$$\frac{\partial^2 G}{\partial \zeta^2} + \frac{1}{2} H \frac{\partial G}{\partial \zeta} = 0 \tag{3.30}$$

this indicates that H is, initially, a Blasius profile and consequently so is

G. Near s' = 0 the appropriate similarity variable depends on whether the spin or the radial flow dominates so these two cases need to be considered separately. This has been done by the authors (ref.24) but as far as the separating efficiency of the cyclone is concerned the scalings described in eqn. (3.26) is usually sufficient. This is because when interest lies in the boundary layer on the top wall, the integrations need only be performed up to the radius of the outer wall of the vortex finder and in the side wall boundary layer analysis, the core flow begins to break down near s' = 0.

The partial differential eqns. (3.27) and (3.28) are expressed in finite difference form according to the following scheme. Firstly writing the equations in terms of $\partial F/\partial \zeta = q$ the derivatives in the s' direction are represented in central difference form. Thus, writing

$$\left.\begin{aligned} P &= q_1 + q_2 \\ T &= G_1 + G_2 \end{aligned}\right\} \qquad (3.31)$$

where the suffices 1 and 2 denote the values at s'_1 and s'_2 ($s'_1 < s'_2$) respectively, eqns. (3.27) and (3.28) become

$$\frac{d^2P}{d\zeta^2} - \frac{1}{2}\frac{dP}{d\zeta}\{\chi_1 \int P\, d\zeta - \chi_2 \int q_2\, d\zeta\} + P(1 - s')\{\chi_2 P - \chi_4 q_2\}$$

$$+ \frac{1}{2} K^2(T^2 - 4)(1 - s') - 2(1 - s') = 0 \qquad (3.32)$$

$$\frac{d^2T}{d\zeta^2} - \frac{1}{2}\frac{dT}{d\zeta}\{\chi_1 \int P\, d\zeta - \chi_2 \int q_2\, d\zeta\} - \frac{s'(1 - s')}{\Delta s'} P\,(2G_2 - T) = 0$$

$$(3.33)$$

Here $\Delta s' = s'_1 - s'_2$ and

$$\left.\begin{aligned} \chi_1 &= \frac{s'}{2} - (1 - s') + \frac{2\, s'(1 - s')}{\Delta s'} \\ \chi_2 &= \frac{4\, s'(1 - s')}{\Delta s'} \\ \chi_3 &= (1 - s')(\frac{1}{2} + \frac{s'}{\Delta s'}) \\ \chi_4 &= \frac{2\, s'(1 - s')}{\Delta s'} \end{aligned}\right\} \qquad (3.34)$$

The coupled ordinary differential equations (3.32) and (3.33) are now written in central difference form. Using a method developed by Merkin (ref.25), the non-linear finite-difference equations, subject to the boundary conditions, are solved using the Newton-Raphson method, to an accuracy of 10^{-6} at each stage. This was always achieved in less than seven iterations. The system of linear equations resulting from the Newton–Raphson scheme was solved by Choleski decomposition into upper and lower triangular matrices, so that the particular form of the equations could be used to keep the computer storage to a minimum and the equations solved as efficiently as possible.

The errors arising from using finite differences in the s' direction were kept to a minimum by integrating from s'_2 to s'_1 say in first one and then two steps and ensuring that the differences between the two solutions were within some prescribed limits (5×10^{-5}) in the majority of calculations performed. In a similar way tests were performed to see if the radial step length could be increased.

As the solutions progressed towards the centre it became evident that the boundary layer was thickening in a manner similar to that reported by Burggraf *et al* (ref.20). Rather than overcome this difficulty by resorting to a distinct change in mesh size at some station ζ it is best to use an exponentially scaled mesh across the boundary layer. This avoids the previously encountered instabilities which were attributed to the coarse mesh in the outer regions of the boundary layer and probably magnified by the scale of the discontinuity in the mesh size.

In order to preserve an almost uniform mesh near the solid boundary it is best to use

$$\zeta = e^{y} - 1 \tag{3.35}$$

and a uniform mesh size in y. Several mesh sizes were used in the numerical calculations and it was found that 0.1 gave results to an accuracy of 10^{-4}. Using this change of variables enabled the inner details of the flow to be maintained and at the same time account taken of the outer structure.

3.4 Comparison of Pohlhausen and finite difference methods

Given U and V at the outer edge of the boundary layer both the Pohlhausen and the finite difference methods may be used. In order to compare the two methods and also results obtained by other authors the mode of operation of the cyclone is taken where nearly all the fluid is withdrawn from the apex, i.e. the hydrocyclone is being used to separate light particles from the fluid. Also, for comparison reasons, the spin velocity is taken to be that of a free vortex. Then

$$U' = 1/s'^2 \quad \text{and} \quad V' = 1/s' \tag{3.36}$$

Fig.11 shows the radial velocity profiles across the boundary layer at various values of s' with K=5 and m=7 as obtained by the Pohlhausen method.

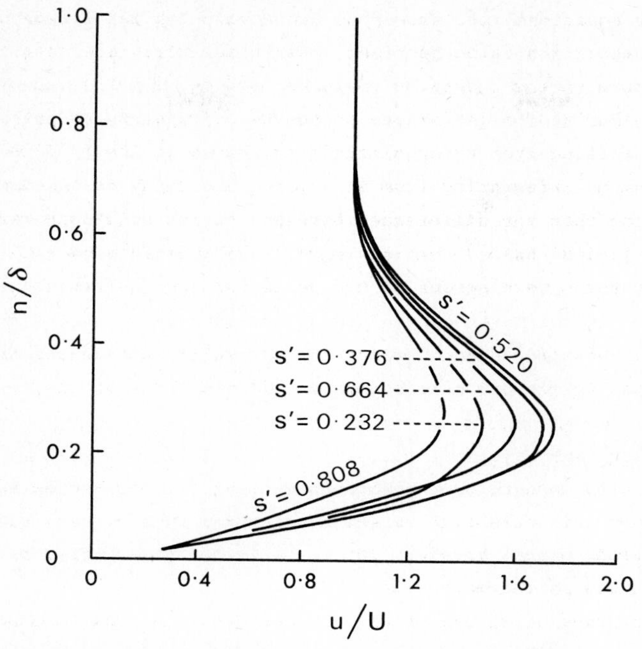

Fig.11 The radial velocity profiles across the boundary layer at various values of s' with K=5 and m=7.

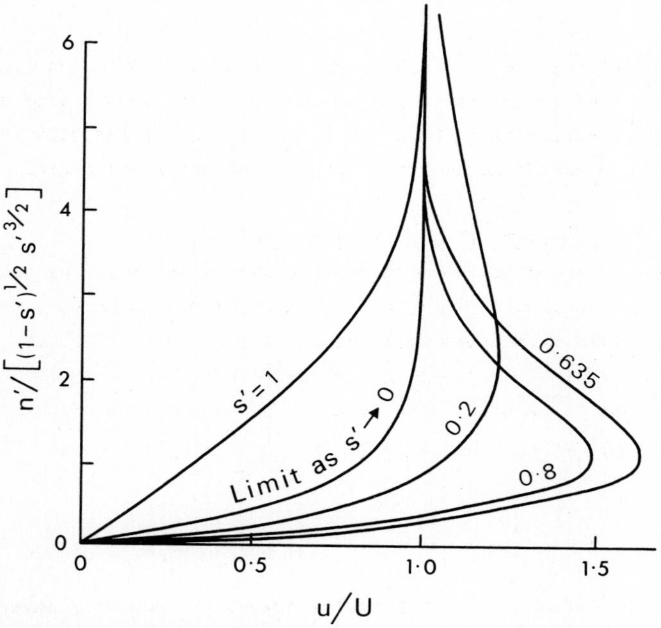

Fig.12 The radial velocity profiles across the boundary layer at various values of s' and K=5. Full finite-difference solution.

As the vertex is approached it is seen that the velocity ratio u/U builds up to a maximum and then starts to decrease. This maximum value is 1.7 and occurs at $s' \simeq 0.5$, $n/\delta = 0.25$. Fig.12 shows the corresponding results for the solution of the full finite difference equations. The same general features exist with the maximum value of $u/U \simeq 1.66$ at $s' \simeq 0.635$, $n/\delta \simeq 0.25$.

Although only one comparison has been made here, many other computations have been performed for a range of values of the parameters and outer flows. All results obtained by the two methods show reasonable agreement. It was also found unnecessary to use such high order boundary conditions on the Pohlhausen method at the outer edge of the boundary layer. Very accurate results could be obtained using no more than the cubic expressions for the profiles and ignoring the pressure boundary condition (3.15).

3.5 Side wall boundary layer

Here a very simple form of the Pohlhausen method is presented but the results obtained are in excellent agreement with the much higher order Pohlhausen method described earlier and the 'exact' finite differnce method.

Here it is assumed that

$$\left.\begin{aligned} u/U &= f_0(\eta) + \lambda\, f_1(\eta) \\ v/V &= g(\eta) \end{aligned}\right\} \tag{3.37}$$

In this case $\sigma^* = \alpha^*$ and for the external flow

$$\left.\begin{aligned} U &= U_0 \left[\frac{s}{s_0}\right]^{-1/2} \\ V &= \frac{V_0}{(R/R_0)} \; \frac{\gamma\left(\frac{4}{5}, \frac{2}{5} L\,(R/R_0)^{5/2}\right)}{\gamma\left(\frac{4}{5}, \frac{2}{5} L\right)} \end{aligned}\right\} \tag{3.38}$$

as obtained from the theoretical analysis presented in the sections (2.2) and (2.3).

The boundary conditions to be satisfied are

$$u = v = w = 0 \qquad \text{at } n = 0 \tag{3.39a}$$

and at the outer edge of the boundary layer

$$\left.\begin{aligned} &u = U(s) \qquad v = V(s) \\ &\text{and} \\ &\partial u/\partial n = \partial v/\partial n = 0 \end{aligned}\right\} \tag{3.39b}$$

Thus the forms chosen for the functions $f_0(\eta)$, $f_1(\eta)$ and $g(\eta)$ are

$$\left.\begin{aligned} f_0(\eta) &= 2\eta - \eta^2 \\ g(\eta) &= 2\eta - \eta^2 \\ f_1(\eta) &= \eta(1-\eta)^2 \end{aligned}\right\} \qquad (3.40)$$

These expressions satisfy the important consistency relations (3.19) and (3.21). Substituting expressions (3.38) and (3.40) into the momentum integral eqns. (3.8) and (3.9) gives

$$\begin{aligned} &\{(I_{00} + 2\lambda_1 J_0 + \lambda_1^2 I_{11}) - (\tfrac{2}{3} + \tfrac{\lambda_1}{12})\}\, \delta' \frac{d\delta'}{ds'} \\ &+ \{2J_0 + 2\lambda_1 I_{11} - \tfrac{1}{12}\}\, \delta'^2 \frac{d\lambda_1}{ds'} - \frac{1}{2}\frac{\delta'^2}{s'}(\tfrac{2}{3} + \tfrac{\lambda_1}{12} - 1) \\ &+ K^2 \delta'^2 V'^2 (1 - I_{00}) = - s'^{1/2}(2 + \lambda_1) \end{aligned} \qquad (3.41)$$

$$\begin{aligned} &\{s'^{-1/2} V'(I_{00} + \lambda_1 I_{01}) - s^{-1/2} V' (\tfrac{2}{3} + \tfrac{\lambda_1}{12})\} \frac{d\delta'}{ds'} \\ &+ \{s'^{-1/2} V' \delta I_{01} - s^{-1/2} V'\delta'/12\} \frac{d\lambda_1}{ds'} \\ &+ s'^{-3/2}\{\tfrac{3}{2}\delta' V' (I_{00} + \lambda_1 I_{01}) - \tfrac{1}{2} V' \delta' (\tfrac{2}{3} + \tfrac{\lambda_1}{12})\} \\ &+ s'^{-1/2}\delta'(I_{00} + \lambda_1 I_{01}) \frac{dV'}{ds'} = -2\frac{V'}{\delta'} \end{aligned} \qquad (3.42)$$

where the additional quantities are non-dimensional variables defined by

$$\left.\begin{aligned} \delta' &= \delta \Big/ \left[\frac{\nu^{1/2} s_0^{1/2}}{(-U_0)^{1/2}} \right] \\ V' &= V / V_0 \end{aligned}\right\} \qquad (3.43)$$

Eqns. (3.41) and (3.42) are two first order simultaneous ordinary differential equations in δ' and λ_1. The boundary layer is assumed to start at $s=s_0$, that is at $s'=1$, $\delta'=0$. Also, the boundary condition (3.18) is applicable. Thus, the initial behaviour is determined and eqns. (3.41) and (3.42) can be integrated numerically using the Runge Kutta Merson method.

For K=4 and L=200 (see eqn.(2.32)), values appropriate to Kelsall's experimental data, Fig.13 shows the distribution of the velocity component parallel to the generators of the cone across the boundary layer at various stations along the wall. It can be seen that the overshoot velocity increases as the vertex is approached. In fact, for sufficiently large values of λ_1, the maximum overshoot velocity is $4\lambda_1/27$ and it occurs about one third of the way across the boundary layer. Figs. 14 and 15 show the variation of λ_1 for various values of the parameters K and L. The value of λ_1 decreases sharply just before the vertex of the hydrocyclone with a consequent reduction in the overshoot velocity; however it is doubtful whether the theory is valid so close to the vertex.

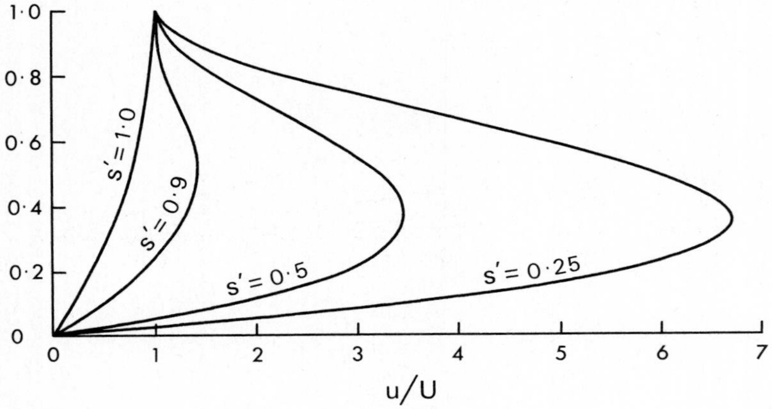

Fig.13 The velocity u as a function of $\zeta \to \eta$ for various values of s'.

Abrahamson, Martin and Wong (ref.26) measured velocities in a long cone hydrocyclone using a pitot-cylinder. They observed that the overshoot velocities near the wall were of the same order of magnitude as those predicted theoretically by the method presented here (ref.9).

The boundary layer thickness is shown in Figs. 16 and 17 as a function of the distance along the wall of the cyclone. With L fixed at 200 the effect of varying the parameter K can be seen in Fig.16. With the velocity component U_0 fixed it can be seen that the effect of increasing the spin velocity is to thicken the boundary layer. However, when the spin velocity is kept fixed, bearing in mind that δ is scaled by $(-U_0)^{1/2}$, it can be seen that the boundary layer thickness at any station is approximately the same. The increased overshoot velocity in the boundary layer caused by the increased spin results in the thickening of the boundary layer. The variation of the boundary layer thickness with the parameter L for a fixed value of K is shown in Fig.17. For increasing L the boundary layer thickens near the vertex of the cyclone simply because the overshoot velocity is maintained by the higher spin that exists there.

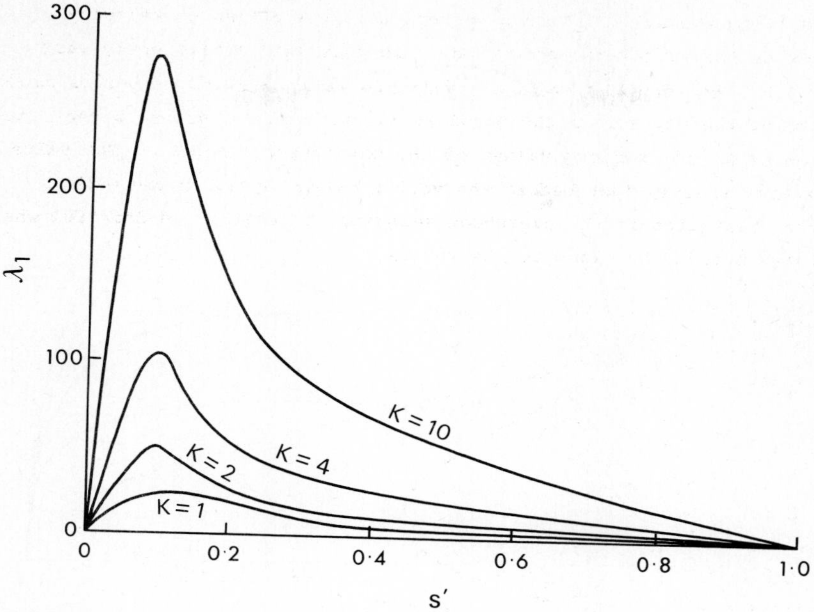

Fig.14 The variation of λ_1 with s' for various K and L=200.

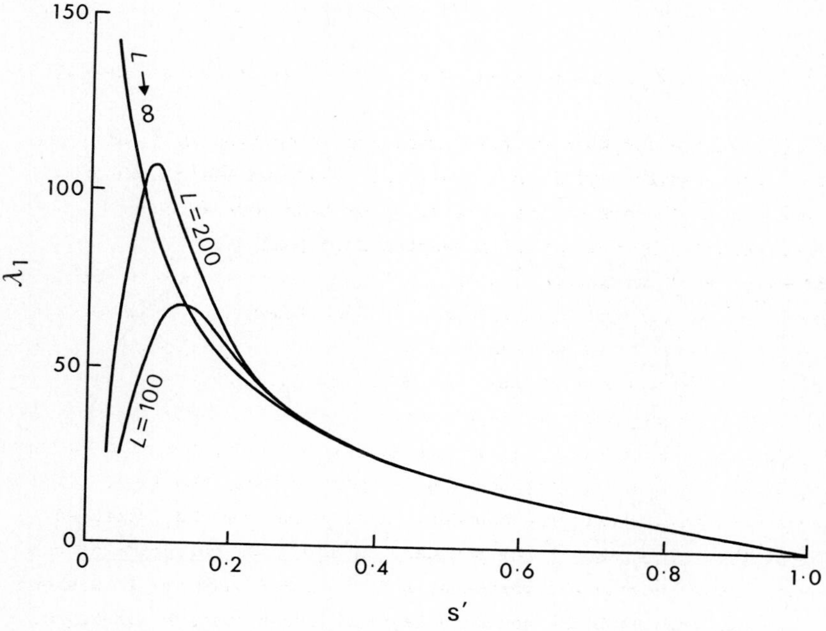

Fig.15 The variation of λ_1 with s' for various L and K=4.

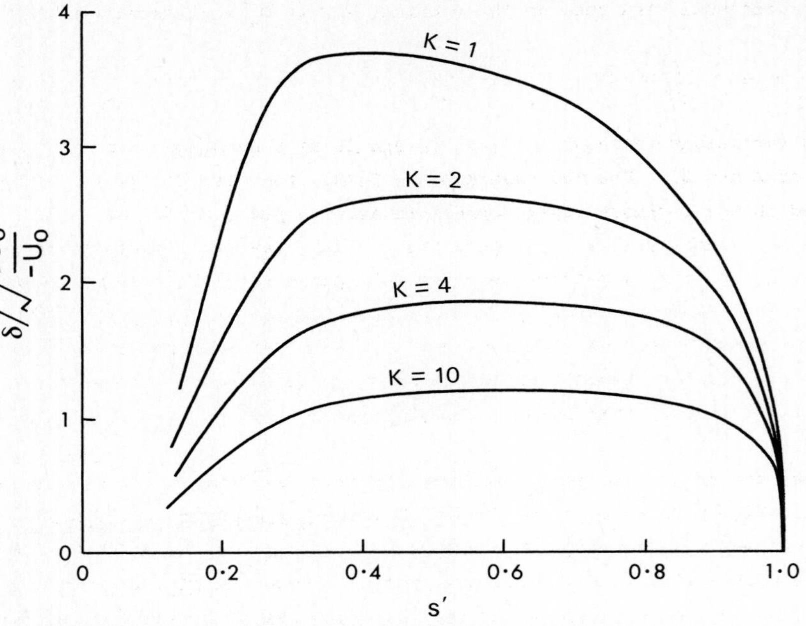

Fig.16 The variation of the boundary layer thickness with s' for various values of K and L=200.

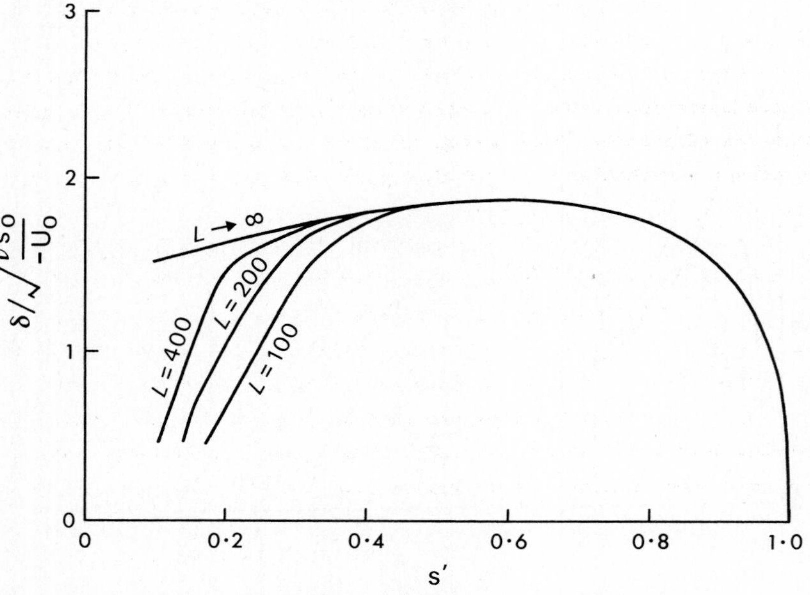

Fig.17 The variation of the boundary layer thickness with s' for various values of L and K=4.

The meridional flux through the boundary layer, Q_s say, is defined as

$$Q_s = s \sin \alpha^* \int_0^\delta u \, dn$$

and its variations is shown in Figs. 18 and 19 as a function of s' for various values of K and L. For the case K=4 and L=200, the maximum value of Q_s attained in the boundary layer represents about 8 per cent of the total flux through the boundary layer. Because the boundary layer thickness is not necessarily accurately predicted by the Pohlhausen method, it might be thought there could be some error in Q_s from the contribution of $Us\delta\sin\alpha^*$. However, the value for the induced flux should be reliable and it can be seen from Fig.13 that this is the main contribution to Q_s owing to the large overshoot velocities. Hence, errors in Q_s from this source are not expected to be significant.

When the spin velocity is due entirely to a free vortex on the axis of the cone, Fig.18 shows that the flux Q_s builds up from entry and finally levels off near the vertex. The effect of allowing for the transition from the free vortex flow to the solid body rotation near the axis for the spin velocity can now be seen as an earlier levelling off of Q_s and a rapid fall as the vertex is approached. This of course is due to the reduced contribution from the overshoot velocity. Fig.19 also shows the variation of Q_s with the parameter K for L=200. Again the effect of the reduction in the overshoot velocity due to the decreasing value of K is shown.

From a practical point of view this is an important aspect of the hydrocyclone operation. When the spin velocity is increased, thus improving the separation efficiency, the induced overshoot velocity also increases and thus improves the mechanism by which dense paticles are transported to the underflow.

Very close to the vertex of the cone the results cannot be regarded as reliable because the validity of the model for the flow external to the boundary layer is questionable in this region. However, Figs. 18 and 19 show a decrease in the value of the flux, over a substantial range of s', as the vertex is approached. This could represent an injection of vorticity into the main body of the flow, a conjecture that is supported by the observations of Rosenweig, Ross and Lewellen (ref.27). Conceivably this represents a contribution to the vorticity distribution function $f(\psi)$ introduced in section (2.2).

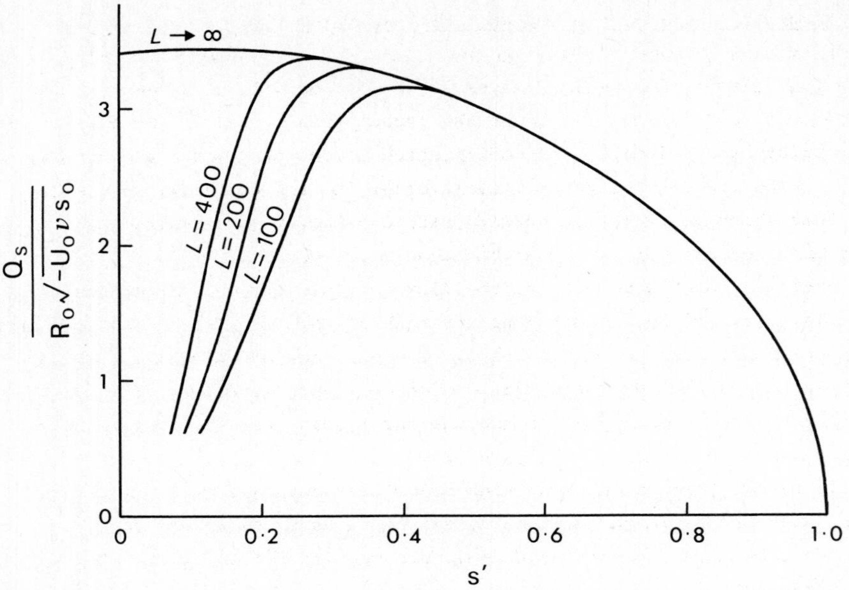

Fig.18 The variation of the meridional flux through the boundary layer as a function of s' for K=4 and various values of L.

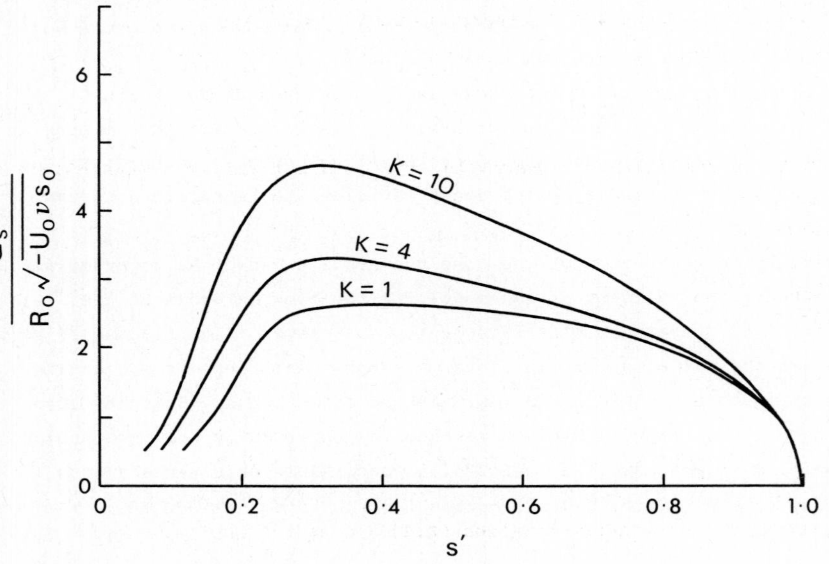

Fig.19 The variation of the meridional flux through the boundary layer as a function of s' for L=200 and various values of K.

3.6 The boundary layer on the top wall of the hydrocyclone

In this case $\sigma^* = \pi/2$. Although eqn. (2.36) gives the spin velocity distribution outside the boundary layer it is not possible to prescribe the radial velocity U. Hence the use of the 'exact' finite difference method is not appropriate and a substantial modification of the Pohlhausen method is required. The presence of the toroidal vortex in the upper region of the hydrocyclone which generates an outward radial velocity just outside the boundary layer on the top wall must be taken into account.

In order to be able to predict the flux of fluid through the boundary layer it is necessary to estimate the strength of the toroidal vortex and this can be done by matching a typical velocity at the edge of the vortex with the velocity in the main body of the hydrocyclone. The parameters which the determine the leakage are obtained and the theory can then be applied to the experimental data given by Kelsall (ref.3).

Again rather than taking such high order polynomials for the velocity profiles, attention will be confined to simple forms since it has been established that more sophistication does not significantly improve the results obtained.

Because of the fundamental change in the nature of the problem it is necessary to reformulate the boundary conditions to be used in conjunction with the eqns. (3.8) and (3.9).

As is usual in the momentum integral methods a solution is sought in similarity form so that

$$u = U\, f(\eta) \qquad v = V\, g(\eta) \tag{3.44}$$

where $\eta = n/\delta(R)$ is the similarity variable, $f(\eta)$ and $g(\eta)$ being dimensionless functions of η.

The meridional flux through the boundary layer, Q_L, is defined by

$$Q_L = R \int_0^{\delta} u\, dz$$

which, using equation (3.44) becomes

$$Q_L = R\, \delta\, U \int_0^{1} f(\eta)\, d\eta = R\, \delta\, U\, C \tag{3.45}$$

Integrating the continuity equation (3.4) across the boundary layer the gives

$$w \Big|_{n=\delta} = -\frac{1}{R}\frac{dQ_L}{dR} + \frac{Q_L}{R\delta C}\frac{d\delta}{dR} \tag{3.46}$$

Introducing parameters D_1, D_2 and D_3, defined by

$$\left.\begin{aligned} D_1 &= C/\int_0^1 f\,(1-g)\,d\eta \\ D_2 &= \frac{1}{C^2}\int_0^1 f^2\,d\eta \\ D_3 &= \int_0^1 (1-g^2)\,d\eta \end{aligned}\right\} \qquad (3.47)$$

eqns. (3.8) and (3.9) become

$$D_2\frac{d}{dR}\left[\frac{Q_L^2}{R\delta}\right] + D_3 V^2\delta - \frac{Q_L}{CR\delta}\frac{dQ_L}{dR} - \frac{Q_L}{C^2}\frac{d}{dR}\left[\frac{Q_L}{R\delta}\right]$$

$$= -\frac{R}{\rho}\tau_R\Big|_{n=0} \qquad (3.48)$$

$$\frac{d}{dR}\left[Q_L V R\right] - D_1 Q_L\frac{d}{dR}\left[V R\right] = \frac{D_1}{\rho} R^2\tau_\theta\Big|_{n=0} \qquad (3.49)$$

In the case of laminar flow the appropriate expressions for the shearing stresses at the wall are

$$\left.\begin{aligned} \tau_R\Big|_{n=0} &= \frac{\mu Q_L}{R\delta^2 C} f'(0) \\ \tau_\theta\Big|_{n=0} &= \frac{\mu V}{\delta} g'(0) \end{aligned}\right\} \qquad (3.50)$$

When the flow is turbulent a modification of Blasius' resistance formula (derived by von Karman (ref.28) and used by Rott and Lewellen (ref.16)) is employed, that is

$$\left.\begin{aligned} \tau_R\Big|_{n=0} &= \frac{0.0225\,\rho\,Q_L V}{CR\delta}\left[\frac{\mu}{\rho V\delta}\right]^{1/4} \\ \tau_\theta\Big|_{n=0} &= 0.0225\,\rho\,V^2\left[\frac{\mu}{\rho V\delta}\right]^{1/4} \end{aligned}\right\} \qquad (3.51)$$

In the case of laminar flow the forms chosen for the profiles are

$$\begin{aligned} f(\eta) &= \beta\,\eta^3 - (1 + 2\beta)\,\eta^2 + (2 + \beta)\,\eta \\ g(\eta) &= 2\eta - \eta^2 \end{aligned} \tag{3.52}$$

where β is a parameter which determines the scale of the maximum velocity in the boundary layer. These expressions satisfy the boundary conditions (3.39). A lower order approximation would produce physically unrealistic velocity profiles.

Using criteria similar to those used in the laminar flow problem, the profiles chosen for the turbulent case are

$$\begin{aligned} f(\eta) &= \eta^{1/7}\,[\gamma + (15/7 - 2\gamma)\,\eta + (\gamma - 8/7)\,\eta^2] \\ g(\eta) &= \eta^{1/7}\,(8 - \eta)/7 \end{aligned} \tag{3.53}$$

where γ plays the role of β for the turbulent case. These agree with the 'one-seventh power velocity distribution law' near the wall (see Schlichting (ref.14)).

Using eqns. (3.52) and (3.53) the coefficients D_1, D_2 and D_3 can be determined from eqns. (3.47) for the laminar and turbulent cases respectively.

It is convenient to make eqns. (3.48) and (3.49) dimensionless in the following way

$$\left.\begin{aligned} v &= V_i v' \\ R &= R_c R' \\ Q_L &= (\mu/\rho)^{\omega/(1+\omega)} (V_i R_c)^{1/(1+\omega)} R_c Q_L' \\ \delta &= (\mu/\rho)^{\omega/(1+\omega)} (V_i R_c)^{-\omega/(1+\omega)} R_c \delta \end{aligned}\right\} \tag{3.54}$$

In the case of laminar flow $\omega=1$ whereas for the turbulent case $\omega=1/4$. Hence the dimensionless forms of the first order ordinary differential equations (3.48) and (3.49) have to be solved for a given external swirling velocity.

Because the boundary layer on the top wall only extends inwards as far as the vortex finder, it is reasonable, and certainly more simple, to take V(R) given by eqn. (2.19).

Thus, eqns. (3.48) and (3.49) can now be written in the form

$$\frac{dQ'_L}{dR'} = D_1 c_1 \left[\frac{R'}{\delta'} \right]^{\omega} \qquad (3.55)$$

and

$$D_2 \frac{d}{dR'} \left[\frac{Q'^2_L}{R'\delta'} \right] + D_3 \frac{\delta'}{R'^2} - \frac{Q'_L}{R'\delta' C} \frac{dQ'_L}{dR'} - \frac{Q'_L}{C^2} \frac{d}{dR'} \left[\frac{Q'_L}{R'\delta'} \right]$$

$$= - \frac{c_2 Q'_L}{R'\delta'} \left[\frac{R'}{\delta'} \right]^{\omega} \qquad (3.56)$$

where

$c_1 = 2$ (laminar)

$c_1 = 0.0225$ (turbulent)

and

$c_2 = 12\ (2 + \beta)/(8 + \beta)$ (laminar)

$c_2 = 0.0225/(\frac{343}{1320} \gamma - \frac{7}{11})$ (turbulent)

The appropriate boundary conditions are

$$Q'_L = \delta' = 0 \quad \text{at} \quad R' = 1 \qquad (3.57)$$

assuming that the boundary layer originates at the outer edge of the hydrocyclone.

Owing to the singular nature of the equations at R'=1, it is necessary to determine analytically the behaviour there before a numerical intergration can be started. Therefore, writing

$$R' = 1 - \epsilon, \quad Q'_L = k\, \epsilon^{\sigma^*_1}, \quad \delta' = h\, \epsilon^{\sigma^*_2} \qquad (3.58)$$

where ϵ is small and h, k, σ^*_1 and σ^*_2 are constants , and by satisfying eqns. (3.55) and (3.56) give

$$\left.\begin{aligned} \sigma^*_1 &= (2 + \omega)/2(1 + \omega) \\ \sigma^*_2 &= 1/2(1 + \omega) \end{aligned}\right\} \qquad (3.59)$$

$$h = \left\{ D_1 c_1 \left(\frac{A\sigma_1^*}{D_1 c_1} + B\sigma_2^* \right)^{1/2} / \left(\sigma_1^* D_3^{\ 1/2} \right) \right\}^{1/(1+\omega)}$$

$$k = - \frac{D_1 c_1}{\sigma_1^*} (1/h)^{\omega} \qquad (3.60)$$

where $A = D_1 c_1 (2D_2 - 1/C - 1/C^2)$ and $B = 1/C^2 - D_2$. The numerical integration of eqns (3.55) and (3.56) can now be carried out with the starting values as given in eqn. (3.58).

The fourth order Runge Kutta Merson method was used and the calculations were performed for a range of values of the parameters β and γ.

For the case of laminar flow the volume flux through the boundary layer as a function of the radial distance for various values of the parameter β is shown in Fig.20. For values of $\beta < -100$ the results are almost identical to those for $\beta = -100$ – indeed, for the range of β values shown it can be seen that the volume flux is fairly insensitive to this parameter. The figure also indicates that increasing the radius of the vortex finder decreases the leakage. However, this would adversely affect the flow pattern and some compromise is necessary. Fig.21 shows the variation of the radial velocity just outside the boundary layer with the radial distance for a range of β values, while Fig.22 shows the boundary layer thickness on the top wall.

In Figs. 23, 24 and 25 similar results are shown for the case of turbulent flow. However, in this case the parameter γ is in the range $-10 \geqslant \gamma \geqslant -25$. For values of $\gamma < -25$ there is no significant change in the volume flux from the case $\gamma=-25$.

For both the laminar and turbulent cases the model used is somewhat restrictive in that the values of β and γ were held constant. This was necessary because the flow just outside the boundary layer was unknown and a more realistic model would allow variations of β and γ with R'. However, fortunately it can be seen from the results that the volume flux and the boundary layer thickness are insensitive to these parameters. Consequently this restriction is not as critical as might first be thought in estimating the leakage but the velocity profiles outside the boundary layer are questionable. Clearly only a rough estimate of β and γ are required and these can be found from an order of magnitude of the velocity in the toroidal vortex. This velocity is estimated from a typical inward radial velocity in the vicinity of the bottom of the toroidal vortex. Using the basic flow model developed in section 2 this can be obtained and estimates of β and γ

found. The volume flux through the boundary layer near the radius of the vortex finder can then be determined.

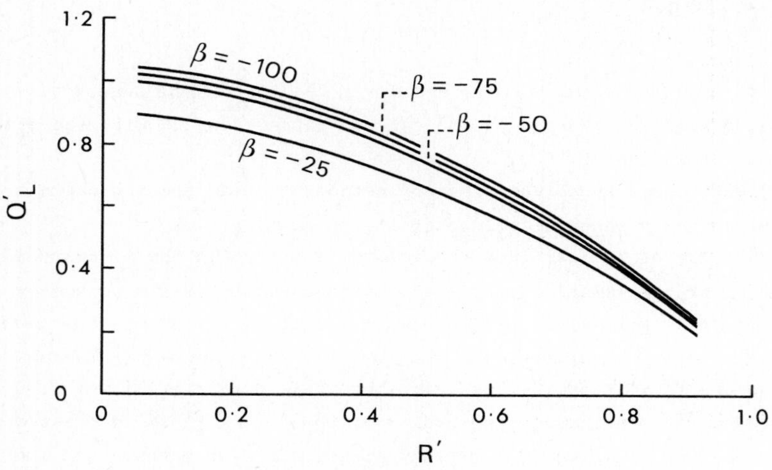

Fig.20 Volume flux through the laminar boundary layer as a function of the radial distance for various values of β.

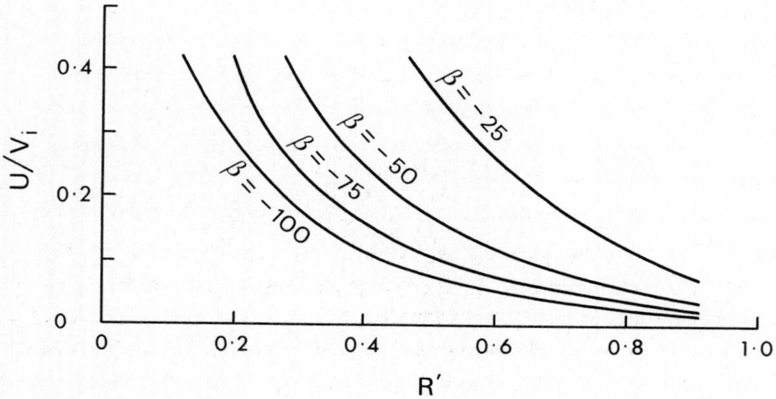

Fig.21 Variation of the radial velocity just outside the laminar boundary layer with the radial distance for a range of values of β.

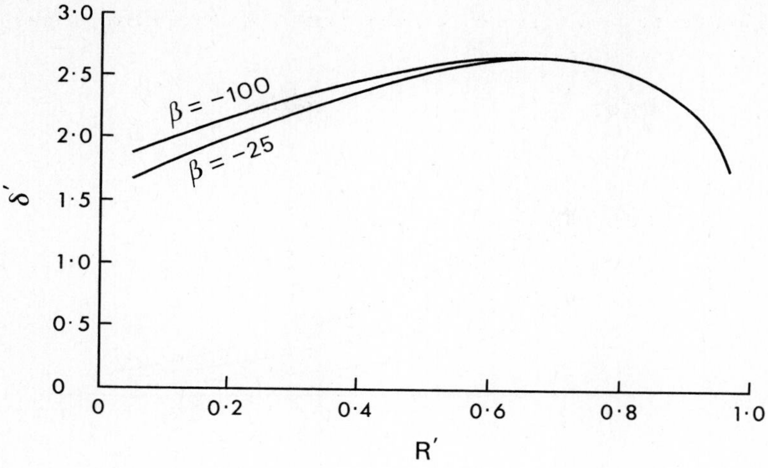

Fig.22 Variation of the laminar boundary thickness on the top wall of the hydrocyclone for various values of β.

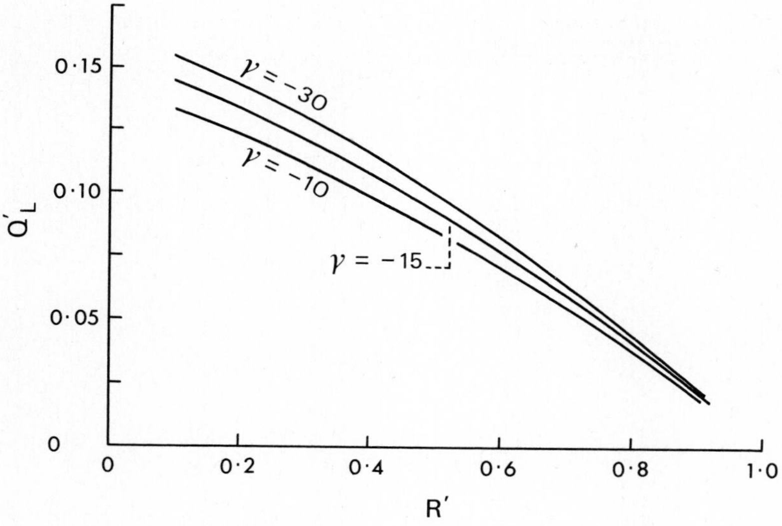

Fig.23 Volume flux through the turbulent boundary layer as a function of the radial distance for various values of γ.

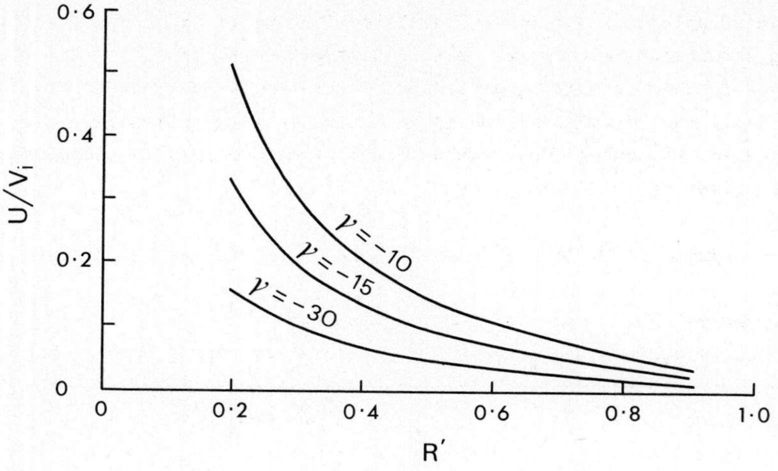

Fig.24 Variation of the radial velocity just outside the turbulent boundary layer with the radial distance for a range of values of γ.

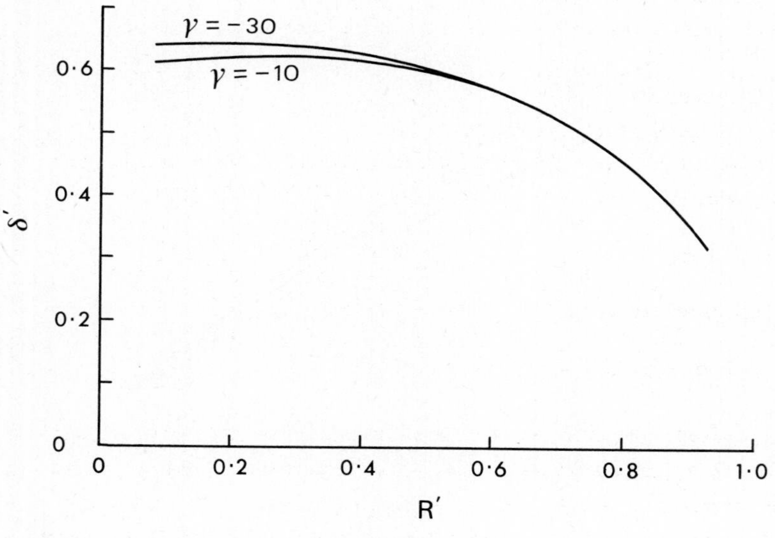

Fig.25 Variation of the boundary layer thickness on the top wall of the hydrocyclone for various values of γ.

As an example of a particular case, the leakage is calculated for the operating conditions relevant to Kelsall's experimental data. The Reynolds number flow in this case is about 10^5 and hence the boundary layer should be laminar. For the laminar flow model an estimate of the parameter β is -100, and at R'=0.25, the outer edge of the vortex finder, $Q'_L \simeq -1$. Hence the leakage is given by

$$-2\pi\ Q_L = -2\pi\ R_c^{3/2} V_i^{1/2}\ \mu^{1/2}\ \rho^{-1/2} Q'_L = 74\ cm^3\ s^{-1}$$

which represents just under 10% of the total flux.

For larger hydrocyclones under normal operating conditions the Reynolds numbers may be so large that the boundary layer is turbulent. For this case the laminar model could still be used with an appropriate eddy viscosity although more reliable results should come from the turbulent model - in which case the leakage is

$$-2\pi\ Q_L = -2\pi\ R_c^{9/5} V_i^{4/5}\ \mu^{1/5}\ \rho^{-1/5} Q'_L$$

Previous workers have suggested that the leakage could be as high as 15%: the present theoretical results are clearly in agreement with this.

4. PARTICLE EFFECTS

4.1 Motion of an isolated particle

For a spherical particle of diameter d placed in a stream of speed $\underline{U}$, the drag $\underline{D}$ is given by

$$D = 3\pi\ \mu\ \underline{U}\ d \tag{4.1}$$

provided the particle Reynolds number R_p defined by

$$\underline{R}_p = |\ \underline{U}\ |\ d\ /\ \nu \tag{4.2}$$

is small compared with unity. It is well known (ref.29) that if this restriction on R_p is not satisfied eqn. (4.1) underestimates the drag and various corrections are available. However, in most cases of interest, R_p is small (ref.1) and eqn. (4.1) is adequate. In practice particle shapes are irregular and under these circumstances it is more convenient to work in terms of an equivalent spherical Stokes' diameter.

The mass of the solid particle is $\rho_s \pi d^3/6$, where ρ_s is the density of the solid, and using eqn. (4.1) it can be seen that a time scale for the acceleration of the particle relative to the fluid is $(\rho_s \pi d^3/6)/(3\pi\mu d)$. A time scale for the flow in a hydrocyclone is a typical time taken for a fluid particle to complete one revolution, namely $2\pi R_c/V_i$. In order that the acceleration time of the particle be negligible it is necessary that

$$\rho_s \, d^2 / 18\mu \ll 2\pi \, R_c / V_i \tag{4.3}$$

This is usually satisfied owing to the dominance of the small value of d, e.g. for Kelsall's apparatus, the inequality (4.3) reduces to the condition that d be small compared with 500μm.

A further complication which arises in the hydrocyclone is that eqn. (4.1) is not strictly valid due to the spinning motion about the axis. Karanfilian and Kotas (ref.30) examined the motion of solid particles through a rotating fluid. They found that for $R_p < 1$ the drag is virtually unaffected by the spinning motion but the particle experienced a lift due to the Corolis acceleration of the liquid particles surrounding the sphere. The lift was of the same order of magnitude as the drag. For the case of the hydrocyclone, the lift force is directed in the azimuthal direction and will produce a minute change in the effective spin velocity of a particle. In the rare cases when R_p is not small, both the lift and drag on the particle are modified in a complicated fashion and reference to the above paper should be made.

When $R_p < 1$ the drag on the particle given by eqn. (4.1) is balanced by the centrifugal force due to the rotation of the liquid. Hence the drift velocity is in the R direction and is given by

$$U_D = (\rho_s - \rho) \frac{d^2 \, q_\lambda^2}{R} \tag{4.4}$$

4.2 Efficiency of the hydrocyclone

Owing to the fact that the performance of a hydrocyclone is virtually unaffected by minor variations in design, the prediction of the effieciency on an *ad hoc* basis has proved fairly satisfactory. There have been very few attempts to estimate theoretically the performance. This is usually assessed by finding the size d_{50} of those particles, of which 50% go to the overflow and 50% to the underflow.

Bradley (ref.31) obtained an expression for d_{50} using the 'equilibrium orbit' concept. The position of zero vertical velocity was taken to coincide with the surface of an imaginary cone whose apex was at the apex of the hydrocyclone and whose base was at the bottom of the vortex finder. Across this surface the radial velocity component was assumed to be constant and equated to the radial settling velocity of the particle of size d_{50} based on

Stokes' Law. Hence an expression for d_{50} was obtained in the form

$$d_{50} = 2.7 \left[\frac{\tan(\alpha^*/2)\ \mu\ (1 - R_f)}{2\ R_c\ Q\ (\rho_s - \rho)} \right]^{1/2} \frac{2.3a}{R_c} \frac{D_i^2}{\alpha} \quad (4.5)$$

Bradley and Pulling (ref.32) modified Bradley's model to take account of the fact that the surface of zero vertical velocity was cylindrical in the upper region of the hydrocyclone. This modification led to the relationship

$$d_{50} = 3.2\ (0.43)^n \frac{D_i^2}{\alpha} \left[\frac{\tan(\alpha^*/2)\ \mu\ (1 - R_f)}{2\ R_c\ Q\ (\rho_s - \rho)} \right]^{1/2} \quad (4.6)$$

In the derivation of the eqns. (4.5) and (4.6) the effect of the leakage of the suspension to the overflow through the boundary layer on the vortex finder had of course been neglected. A further assumption was that the particles instantaneously reached their equlibrium positions, that is, where the radial settling velocity balances the inward fluid velocity. Bradley and Pulling suggested that the inward radial velocity had the required variation on the cone surface. The constant substituted for the radius of the imaginary cone was chosen to give the correct form of the correlation, a change in the constant simply altering the numerical factor in eqn. (4.6).

Lilgé (ref.33) went further on these lines. Using experimental data he took a linear variation of inward radial velocity with height on the surface of the imaginary cone. Making assumptions similar to those made by Bradley and introducing a 'cone force equation' applicable to particles of any size and shape, he was able to obtain an expression for d_{50}.

An expression akin to eqns. (4.5) and (4.6) was obtained by Tarjan (ref.34) using arguments and assumptions similar to those used by Bradley, and Bradley and Pulling.

Rietema has estimated the efficiency of a hydrocyclone by considering a particle injected into a hydrocyclone along the centre line of the inlet pipe. In order to obtain the trajectory of such a particle it was assumed that both the axial and radial components of velocity of the fluid were constant in the separating zone. It can be seen from Kelsall's (ref.3) experimental work that the approximations regarding the velocity components are not strictly valid. Furthermore, the suspension after entering the hydrocyclone does not remain in the tight band defined by the area of the inlet but spreads out. However, the advantage of this model is that is does not rely on the particles instanteously reaching their equilibrium positions.

Huyet *et al.* (ref.35) carried out a survey of working hydrocyclones of widely differing specification. They showed that the performance varied greatly from hydrocyclone to hydrocyclone indicating the difficulty in obtaining a unifying theory.

It will be assumed that the concentration of the solid phase is low so that any particle-particle interactions within the core of the hydrocyclone may be neglected. Obviously when the particles have been separated they will be in a very thin boundary layer on the side wall of the hydrocyclone and here the concentration may be large and the particle-particle interactions could become significant. This point will be discussed later but at the moment the motion of a single small dense solid particle within the core flow of the hydrocyclone will be considered.

As can be seen from Fig.4a the fluid velocity has a component radially inwards so there exists a surface on which a particle of given size will be in equilibrium (as regards the horizontal motion). This surface is commonly referred to as the equilibrium line and is a very important concept when considering the efficiency of the hydrocyclone.

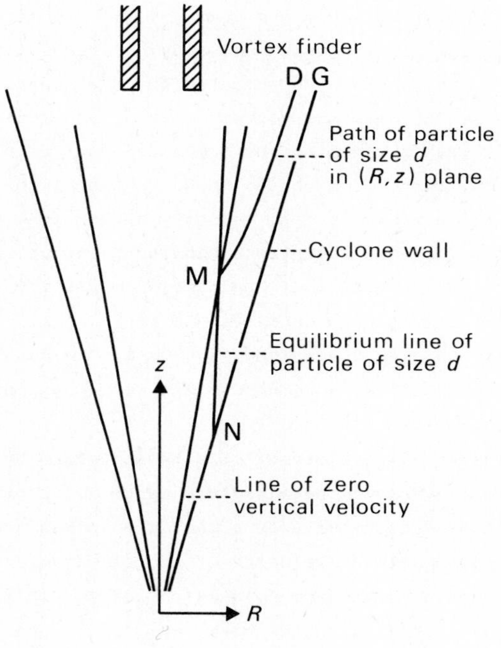

Fig.26 Coordinate system and a schematic representation of the critical path of a particle of size d

With reference to Fig.26 consider a particle which reaches its equilibrium line at the point M where this equilibrium line crosses the line of zero vertical velocity. The trajectory of this particle is determined by integrating back along its path from M until a region is reached where the suspension may be regarded as homogeneous, the point D say. All particles of this size in the annular region between D and G therefore approach their equlibrium line between M and N. Since they are entrained in the downward moving fluid and cannot cross their equilibrium line MN, they are collected in the boundary layer between G and N. The particles thus captured are discharged at the underflow through the boundary layer. This is the only way that a particle can be removed at the underflow.

The remainder of the particles of this size move towards their equilibrium line above M. Clearly they cross the line of zero vertical velocity above M and are thus entrained in the upward moving fluid and are eventually carried through the overflow. The efficiency of the hydrocyclone for this particular size is therefore that fraction of the total flux which passes through the annular region between D and G since the suspension is regarded as homogeneous at this level.

Consider now a particle of 'Stokes' diameter d in the liquid. Balancing the inward radial velocity with the outward drift of the particle, the equilibrium line is given by

$$\frac{B\ R^{3/2}}{z^2} = (\rho_s - \rho)\frac{d^2\ q_\lambda^2}{18\mu R} \tag{4.7}$$

using eqns. (2.15) and (4.4). Also using eqn. (2.14), the line of zero vertical velocity is

$$R = 3\alpha^* z / 5 \tag{4.8}$$

These curves intersect at the point P.

Using eqns. (2.14), (2.15) and (4.4) the trajectory of the particle is determined by the differential equation

$$\frac{dR'}{dz'} = \frac{-R'^{3/2} z'^2 + d^2 q_\lambda'^2/(\lambda^{*2} R')}{R'^{-1/2}(3 - 5R'/z')/2} \tag{4.9}$$

where $q'_\lambda = q_\lambda/V_i$, and the parameter λ^* is defined by

$$\lambda^{*2} = \frac{18\mu B\alpha^{*9/2}\ \ell^{5/2}}{(\rho_s - \rho)\ V_i^2 R_o^2} \tag{4.10}$$

The initial slope of the trajectory at P is given by

$$\frac{dR'}{dz'} = \{15 - \frac{6}{q'_\lambda}\frac{dq'_\lambda}{dR'} + [(15 - \frac{6}{q'_\lambda}\frac{dq'_\lambda}{dR'})^2 - 180]^{1/2}\} / 25 \tag{4.11}$$

Eqn. (4.9) is integrated numerically from the point M to D, the level of the end of the vortex finder, at which the suspension may be regerded as homogeneous. The ratio of the the stream function at D to its maximum value can be determined from eqn. (2.13) and is $E_m(d)$ where

$$E_m(d) = \frac{(\alpha^* \ell/a)^{3/2}}{1 - a/\alpha^* \ell} R'^{3/2}_D (1 - R'_D) \tag{4.12}$$

and R'_D is the value of R' at D. This, of course, is the proportion of particles of size d which are removed in the main body of a hydrocyclone specified by the parameters λ^* and $\alpha^* \ell/\alpha$.

It is now necessary to consider the leakage of the particles through the boundary layer on the top wall of the hydrocyclone. Fluid is entrained into this boundary layer from the external flow over the whole of its surface. In view of the nature of the external flow it seems reasonable to assume that the distribution of particles in this entrained fluid is homogeneous.

Thus the separating efficiency of the flow within the boundary layer can be determined by examining the path followed by the particle of a given size entering the boundary layer. If the particle remains within the boundary layer then it is taken that it leaks to the overflow otherwise it enters the main body of the hydrocyclone and these particles must be accounted for when calculating the distribution of particle sizes.

The equation of the particle path is given by

$$\frac{dz}{dR} = w / [u + \frac{(\rho_s - \rho) d^2 v^2}{18\mu R}] \tag{4.13}$$

where the velocity components u and v are given by eqns. (3.44). Using the continuity equation, w is given by

$$w = -\frac{\delta}{R}\frac{d}{dR}(RU)\int_0^\eta f(\eta)\, d\eta + U\frac{d\delta}{dR}\{\eta f(\eta) - \int_0^\eta f(\eta)\, d\eta\} \tag{4.14}$$

It is more convenient to work in the η, R plane where the particle path, in non-dimensional form, is given by

$$\frac{d\eta}{dR} = \frac{\frac{Q'_L}{\delta'}\frac{d\delta'}{dR'}\,\eta\, f(\eta) \;-\; \frac{dQ'_L}{dR'}\int_0^{\eta} f(\eta)\, d\eta}{Q'_L\, f(\eta) + \frac{(\rho_s - \rho)\, d^2 V_i}{18\mu R_o}\, g^2(\eta)\delta' \int_0^1 f(\eta)\, d\eta} - \frac{\eta}{\delta'}\frac{d\delta'}{dR'} \tag{4.15}$$

For a particle of size d entering the boundary layer at position R' this equation can be integrated to determine the trajectory of the particle. Hence the minimum value of R, equal to R_m, can be determined consistent with the particle escaping from the boundary layer. Particles of this size entering the boundary layer in the region $R < R_m$ are lost to the overflow. Hence the proportion of the particles of size d which leak to the overflow is

$$\frac{Q_L(a) - Q_L(R_m)}{Q_L(a)} = 1 - E_b(d) \tag{4.16}$$

where $E_b(d)$ may be regarded as the separating efficiency of this region.

In working out the overall efficiency it must be remembered that if $D_o(d)$ is the distribution of particles in the fluid which enters the hydrocyclone, then the homogeneous distribution referred to in the preceding analysis is D_h where

$$D_h = \frac{D_o}{1 - E_b\, Q_L(a)/Q} \tag{4.17}$$

This result follows from the conservation of solid particles. The overall efficiency E(d) of the hydrocyclone is therefore given by

$$E(d) = \frac{E_m(d)\, (Q - Q_L(a))}{Q - E_b(d)\, Q_L(a)} \tag{4.18}$$

Fig.27 shows the efficiency E_m as a function of d/λ^* for various values of the parameter $\alpha^* \ell/a$. For increasing $\alpha^* \ell/a$ the performance improves for a given Q but clearly the feed pressure must be increased to maintain this flux.

Fig.28 shows the variation of E_b with the parameter $d\, \{(\rho_s-\rho)V_i/(18\mu R_o)\}^{1/2}$ for various values of the parameter β. The value a/R_o in this figure is 1/4 but variations of this parameter between 1/6 and 1/3 produce only slight variations in E_b, of the order of ± 5%. For a given hydrocyclone decreasing the value of a increases the leakage but only slightly.

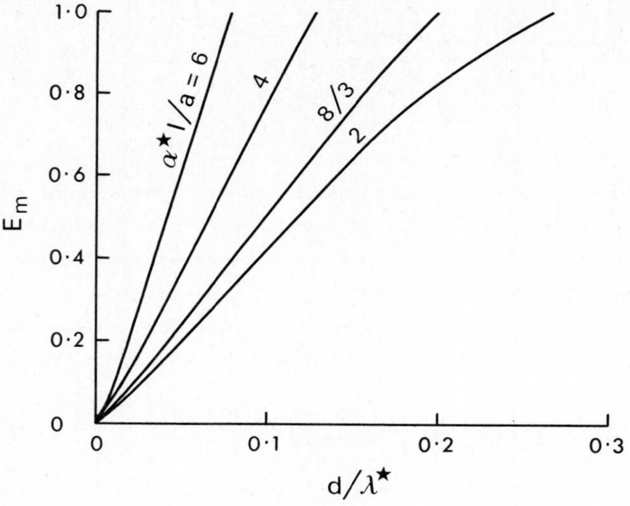

Fig.27 The efficiency, E_m ,as a function of d/λ^* for various values of the parameter $\alpha^* \ell/a$.

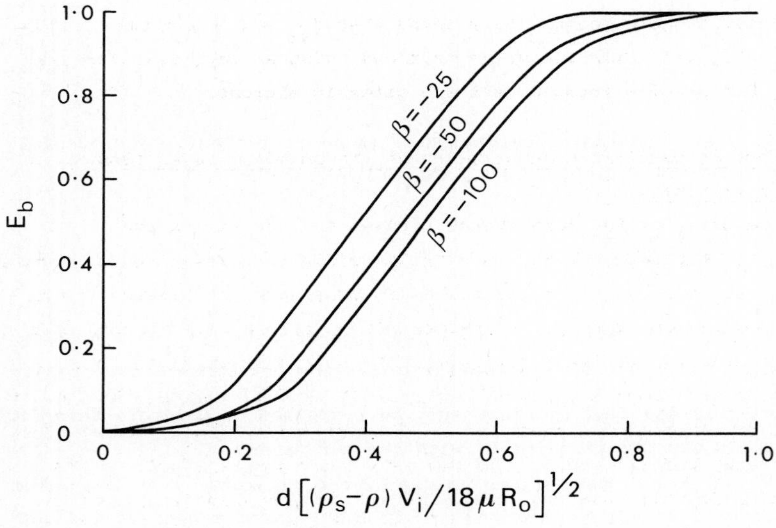

Fig.28 The efficiency, E_b , as a function of $[(\rho s - \rho)V_i/(18\mu R_o)]^{\frac{1}{2}}d$ for various values of the parameter β

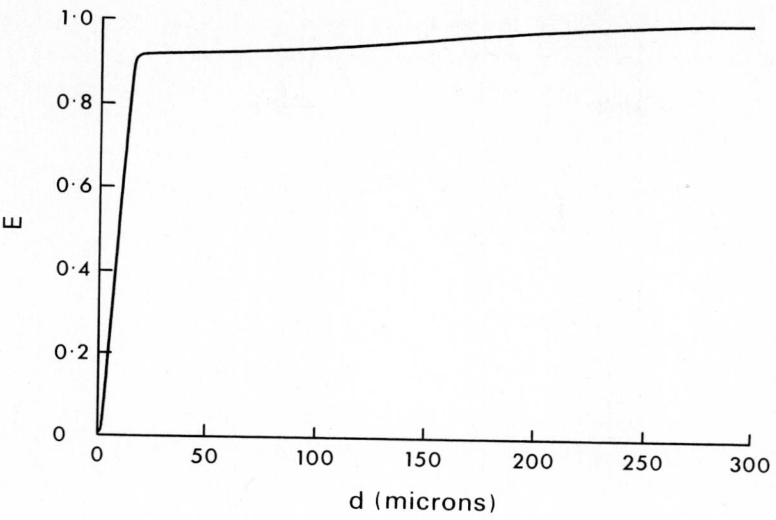

Fig.29 The calculated efficiency curve for Kelsall's hydrocyclone

As an example of the way the theory may be used the parameters are chosen appropriate to the experimental data of Kelsall. Fig.29 shows the efficiency curve calculated on this basis. It is found that d_{50} = 8.6 microns. This agrees reasonably well with the value obtained using an empirical formula; for example, Dahlstrom's formula (ref.37) gives 10 microns.

4.3 The effect of particle concentration on the boundary layer flow – modified viscosity

In the majority of the theoretical studies made so far on the hydrocyclone, it has been assumed that the particle concentration has remained sufficiently low to leave the fluid mechanics unaffected. However, in many industrial applications a very high degree of separation of particles from the injected suspension is desired, giving rise to a large particle concentration within the side wall boundary layer. This will severely hinder the flow even to the point of blocking the underflow. The present analysis follows the work of Laverack (ref.12) and is aimed at determining quantitatively the effect on the side wall boundary layer flow of the particles within the boundary layer. The need to study the flow near the hydrocyclone wall has already been pointed out by Hundertmark (ref.37) who notes that the large concentration of particles there will cause the effective viscosity of the suspension near the wall to be incresased. Every, Hughes and Koch (ref.38) have determined experimentally the viscosity of a coke diesel slurry, as a function of concentration by weight of the coke. The results are presented in Fig.30.

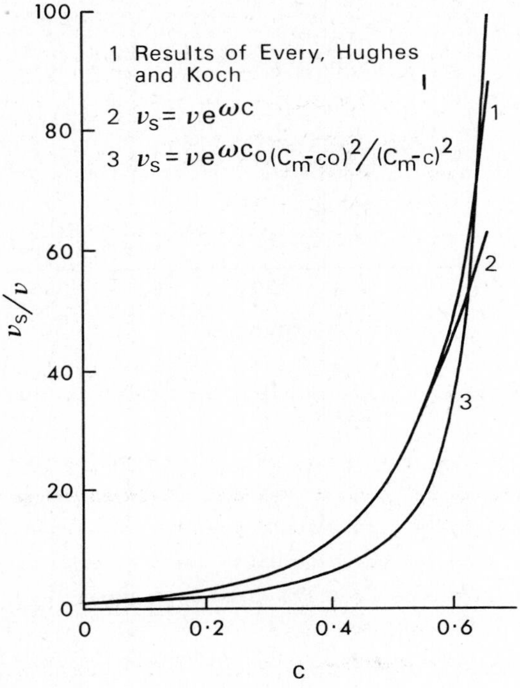

Fig.30 The effective kinematic viscosity of a coke/diesel slurry as a function of particle concentration

Also shown are the two curves

$$\nu_s = \nu \exp(\omega c) \tag{4.19a}$$

$$\nu_s = \nu \,(c_m - c_i)^2 \exp(\omega c_i)/(c_m - c)^2 \,. \tag{4.19b}$$

The former is exponential in c while the latter is algebraic. The curves have been fitted to the experimental results for use in the boundary layer equations, assuming the suspension in a typical hydrocyclone behaves in a similar fashion. Obviously, in a particular case, the viscosity concentration relation depends on the materials to be separated. However, the functions given in eqn. (4.19) are expected to provide results from which general conclusions may be drawn. The effective kinematic viscosity of the suspension in the boundary layer is ν_s , and c is the concentration by weight of the solid particles. The constant ω is chosen so that ν_s from eqn. (4.19a) and the experimental results have the same value at c=0.5, while c_m is the maximum possible concentration by weight within the boundary layer for which a flow can exist.

An alternative to eqn. (4.19a) is eqn. (4.19b) chosen to allow for the fact that the viscosity of the suspension may become singular as $c \to c_m$.

Writing $\nu_s = \nu\ f(c)$ where f(c) corresponds to the function in eqn. (4.19a) or (4.19b) and noting that c is averaged across the boundary layer, then the equations of motion within the boundary layer (eqns. (3.1)-(3.3)) become

$$u \frac{\partial u}{\partial s} + w \frac{\partial u}{\partial n} - \frac{v^2}{s} = U \frac{dU}{ds} - \frac{V^2}{s} + \nu\ f(c) \frac{\partial^2 u}{\partial n^2} \qquad (4.20)$$

$$u \frac{\partial}{\partial s} (sv) + w \frac{\partial}{\partial n} (sv) = \nu\ f(c) \frac{\partial^2}{\partial n^2} (sv) \qquad (4.21)$$

$$\frac{v^2}{s} \cot \alpha^* = - \frac{1}{\rho} \frac{\partial p}{\partial n} \qquad (4.22)$$

The boundary conditions are given by eqns. (3.5).

Before determining the effect of those particles flowing into the boundary layer on the particle concentration within the boundary layer, it is necessary to discover which particles reach the side wall boundary layer and which are removed from the hydrocyclone through the overflow.

With refence to Fig.31, consider a particle of diameter d that reaches the point M where the line of zero vertical velocity crosses the equilibrium line. The trajectory of this particle is determined by integrating back along its path from M until a level is reached where the suspension may be regarded as homogeneous. A suitable level to choose is that of the vortex finder. Thus the particles that reach the point M leave the homogeneous region from the point D. All particles of this size distributed between D and G approach their equilibrium line between M and N, and since they are entrained in the downward moving fluid they enter the boundary layer between G and N. The remainder of the particles of this size are carried to the overflow since their equilibrium line lies in the region of upward moving fluid. Hence only those particles that pass between D and G can enter the boundary layer flow and therefore, assuming no particles return to the main flow in the hydrocyclone from the boundary layer, the mass of particles within the boundary layer flow remains constant between the point N, called the 'cut-off point', and the vertex of the hydrocyclone.

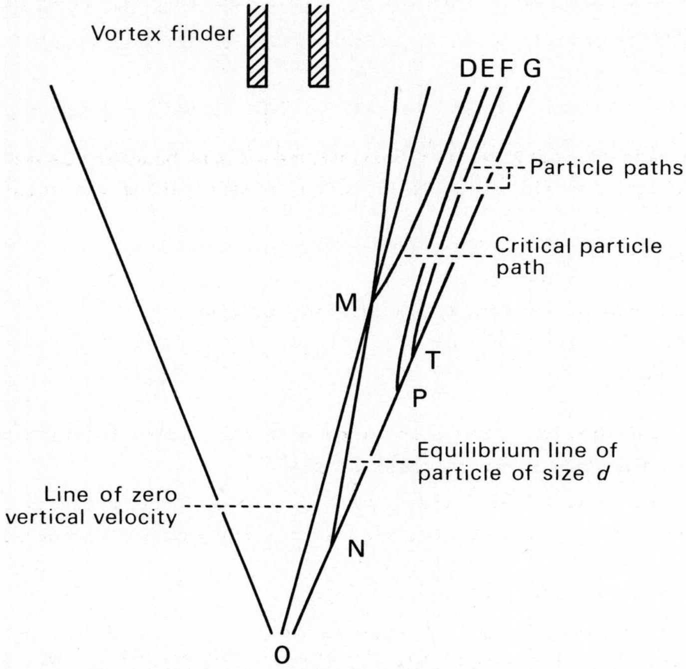

Fig.31 A schematic representation of the trajectories of a particle of size d

Consider now the neighbouring particle paths EP and FT shown in Fig.31. In using the inviscid model for the main flow in the hydrocyclone it is assumed that the boundary layers on the hydrocyclone walls are of negligible thickness and therefore the line GTPN is taken to be the outer edge of the boundary layer rather than the hydrocyclone wall. The meridional volume flux between E and F is given by

$$- (\psi_F - \psi_E) \tag{4.23}$$

where ψ is the stream function. The negative sign indicates that the fluid between D and G is moving towards the vertex of the hydrocyclone.

The mass of particles that pass between E and F can be written as

$$- \xi_0 (\psi_F - \psi_E) \tag{4.24}$$

where ξ_0 is the mass of particles per unit volume of the suspension at this level. The particles between E and F enter the boundary layer between T and P.

The increase in the mass of particles within the boundary layer between T and P can be written

$$Q_P \xi_P - Q_T \xi_T \tag{4.25}$$

where ξ is the mass of particles per unit volume of the boundary layer.

Taking the limit as the distance EF tends to zero yields the equation

$$d(Qs\xi) = - \xi_0 \, d\psi \tag{4.26}$$

This equation can now be integrated immediately to give

$$\xi = - \xi_0 \, \psi / Q_S \tag{4.27}$$

The constant of integration being zero since both the stream function and the flux through the boundary layer are zero at $R=R_O$.

Since ξ is the mass of particles per unit volume of the boundary layer, then the mass concentration c of particles within the boundary layer is

$$c = \xi / [\xi + \rho\{1 - (\xi/\rho_S)\}] \tag{4.28}$$

Eqns. (4.27) and (4.28) determine the effect of those particles that enter the boundary layer on the particle concentration therein, and hence on the overall viscosity of the suspension in the boundary layer as given in eqns. (4.19).

To solve eqn. (4.27) up to a particular station, T say, (see Fig.31), the value of ψ at F is required. In particular, from eqn. (2.13) the coordinates (R_*, ℓ) of the point F are required, where R_* is the value of R at the level $z=\ell$. It is necessary therefore to integrate from T along the particle path to the level of the vortex finder using eqn. (4.9) with, for simplicity, q_λ given by eqn. (2.19).

The momentum integral equations are obtained as before, the result being

$$\frac{\partial}{\partial s} \int_0^\delta s \, u^2 \, dn \; - \; U \frac{\partial}{\partial s} \int_0^\delta s \, u \, dn \; - \; \int_0^\delta v^2 \, dn$$

$$= (s \, U \frac{dU}{ds} - V^2) \, \delta \; - \; \nu \, f(c) \left. \frac{\partial u}{\partial n} \right|_0 \tag{4.29}$$

$$\frac{1}{s^2}\frac{\partial}{\partial s}\int_0^{\delta} s^2 u\, v\, dn - \frac{V}{s}\frac{\partial}{\partial s}\int_0^{\delta} s\, u\, dn$$

$$= -\nu\, f(c)\left.\frac{\partial v}{\partial n}\right|_0 \qquad (4.30)$$

Again, the Pohlhausen method will be used to solve the momentum integral equations. The effect of the streaming motion parallel to the generators of the hydrocyclone is to make the flux Q_s very sensitive to the boundary layer thickness. A high order of resolution in the velocity profiles near the outer edge of the boundary layer is therefore required so that a realistic value is obtained for the boundary layer thickness and hence for the flux Q_s since this value is used in determining the particle concentration within the boundary layer. Consequently, the forms chosen for u and v are given by eqns. (3.10), (3.12) and (3.22). The momentum integral eqns. (4.29) and (4.30) can now be written in non-dimensional form

$$\frac{d}{ds'}\{\delta'\int_0^1 (f_0 + \lambda_1 f_1 + \lambda_2 f_2)(f_0 + \lambda_1 f_1 + \lambda_2 f_2 - 1)\, d\eta\}$$

$$- \frac{\delta'}{2s'}\int_0^1 (f_0 + \lambda_1 f_1 + \lambda_2 f_2 - 1)\, d\eta \; + \; \delta' K^2\, v'^2 \int_0^1 (1 - g^2)\, d\eta$$

$$= - \frac{f(c)}{\delta'}\, s'^{1/2}\, (f_0'(0) + \lambda_1 f_1'(0)) \qquad (4.31)$$

$$\frac{1}{s'^2}\frac{d}{ds'}\{\; s'^{3/2}\, v'\delta' \int_0^1 (f_0 + \lambda_1 f_1 + \lambda_2 f_2)(g - 1)\, d\eta\}$$

$$= - \frac{f(c)\; v'\; g'(0)}{\delta'} \qquad (4.32)$$

In non-dimensional form eqns. (4.27), (4.28) and (2.19) become respectively

$$\xi' = \frac{H\, R_*'^{3/2}(1 - R_*')}{s'^{1/2}\delta'\,(I_0 + \lambda_1 I_1 + \lambda_2 I_2)} \qquad (4.33)$$

$$c = \frac{\xi'}{\xi' + \rho[1 - (\xi'/\rho_s)]} \qquad (4.34)$$

where

$$\xi' = \xi/\xi_0 ; \qquad R_*' = R_*/R_C ; \tag{4.35}$$

$$H = \frac{1}{R_C}\left[\frac{1}{-U_0 \nu s_0}\right]^{1/2} \frac{Q\, m^{5/2}}{2\pi(m-1)} \tag{4.36}$$

and $m_c = R_o/R$ (4.37)

Hence the two boundary layer equations (4.31) and (4.32) subject to the boundary conditions (3.15), (3.17) and (3.18) have to be solved, along with eqns. (4.33), (4.34) and (4.35) in order to determine the concentration of particles within the boundary layer.

Eqn. (4.35) is a first order ordinary differential equation in R' which is integrated using the Runge Kutta Merson method from the outer edge of the boundary layer to the level of the vortex finder in order to find the value of R'_*. The governing equations listed above can then be solved simulateously for δ', λ_1, λ_2, ξ' and c.

The initial concentration of particles within the boundary layer should be equal to the homogeneous concentration in the main flow of the hydrocyclone at the level of the inlet. That is

$$c_0 = \frac{\xi_0}{\xi_0 + \rho[1 - (\xi_0/\rho_s)]} \quad \text{at } s'=1 \tag{4.38}$$

Thus the initial behaviour is determined and eqns. (4.31) and (4.32) are integrated numerically using the Runge Kutta Merson method.

Results are presented corresponding to the operating conditions of Kelsall's hydrocyclone (ref.3) for particles of Stokes' diameter $d=0.1\lambda^*$ which for the case of quartz in water represents about 7.5 microns. Computations were performed for a range of different feed concentrations. The concentration c is plotted as a function of R in Fig.32, and Fig.33 shows the meridional volume flux Q_s as a function of R. Concentrating on the results for a particular feed concentration (0.05 say), the growth of Q_s is seen to have the effect of keeping the particle concentration relatively constant. However, the effect of the spin velocity in the main flow becoming solid body rotation is to cause substantially more fluid to return to the main flow from the boundary layer than if the spin had remained as a free vortex. Consequently the particle concentration increases rapidly as can be seen in Fig.32. The points A, B, and C in these figures represent the cut-off points when the feed concentrations are 5%, 15% and 25% respectively.

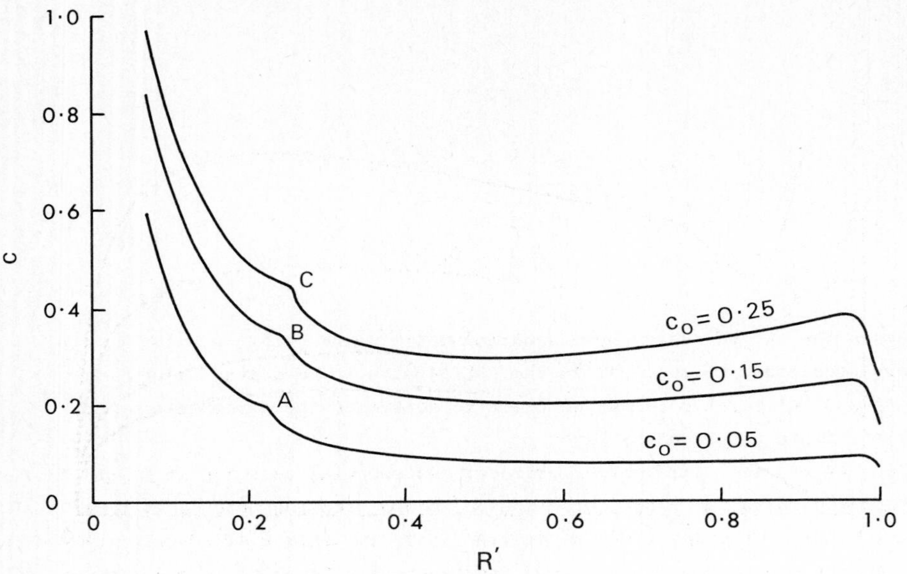

Fig.32 The particle concentration as a function of R' for d/λ^*=0.1 and various values of c_0

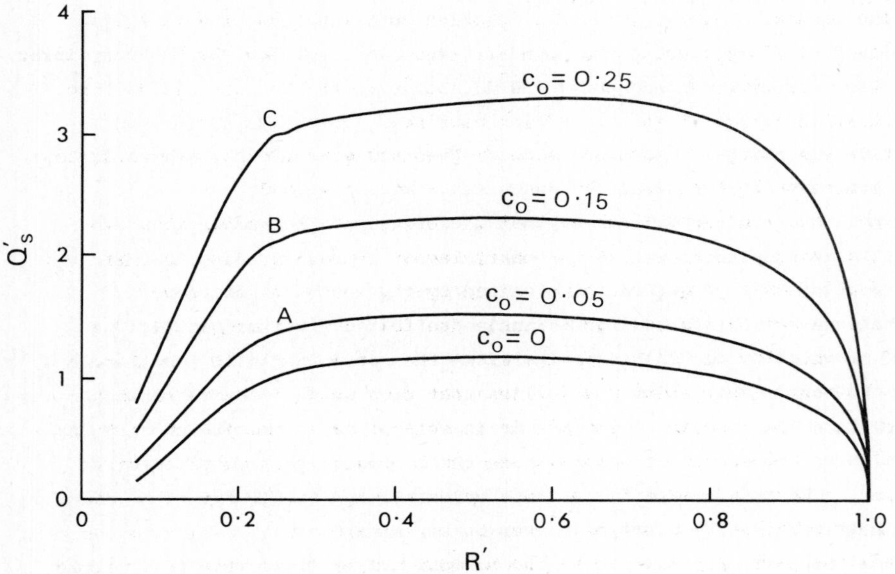

Fig.33 The meridional flux through the boundary layer as a function of R' for d/λ^*=0.1 and various values of c_0

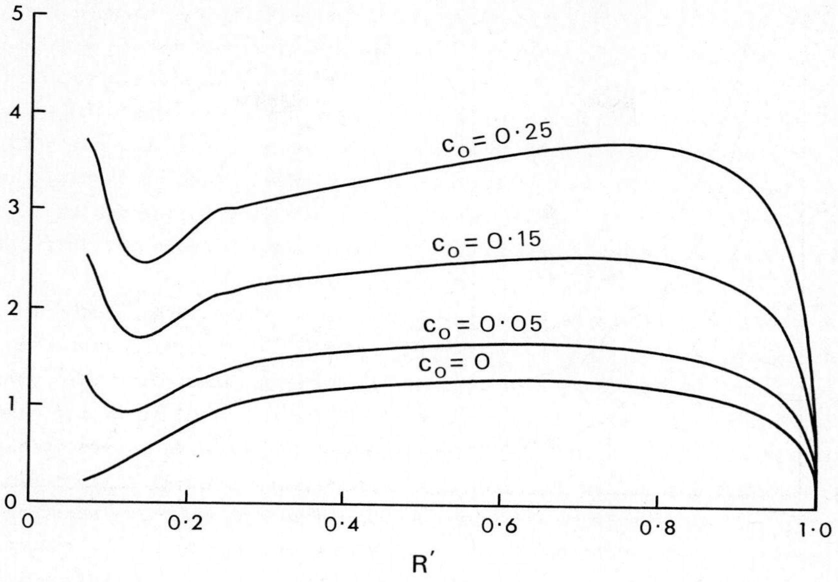

Fig.34 The boundary layer thickness as a function R' for $d/\lambda^*=0.1$ and various values of c_0

The boundary layer thickness is plotted as a function of R in Fig.34. The effect of fluid leaving the boundary layer is to reduce the boundary layer thickness. However, since particles do not leave the boundary layer, the concentration increases in this region causing a rapid increase in the effective viscosity of the suspension. Eventually, therefore, the boundary layer starts to thicken again.

The results clearly indicate that the effect of increasing the feed concentration is to thicken the boundary layer, thus increasing the flux Q_s. Now the boundary layer thickness obtained by the use of a Pohlhausen's technique cannot be regarded as reliable due to the arbitrariness in the choice of velocity profiles, and therefore the dependence of Q_s, and hence c, on the boundary layer thickness implies that care needs to be taken when interpreting the results. In fact it is reasonable to compare the results for different feed concentrations since the boundary layer structures are expected to be similar for the various values of c_o, though the results for each individual feed concentration can be misleading.

All the particles that enter the boundary layer do so between the inlet and the cut-off point and therefore they would all be collected at the underflow if the amount of the suspension within the boundary layer at the level of the cut-off point were removed from the hydrocyclone through the underflow. Of course, it would not be necessary to remove the flow rate at this level. However, Fig.34 shows that the boundary layer structures for the various feed concentrations are similar and therefore it is possible to draw general conclusions from a consideration of the conditions at the cut-off point. Hence with reference to Fig.33, it can be seen that increasing the feed concentration has the effect of increasing the flux Q_s through the boundary layer at the level of the cut-off point and therefore the underflow needs to be increased to cope with the extra flow rate.

As an example it can be seen from Fig.33 that if the feed concentration is increased from 5% to 15% the value of Q_s at the cut-off point increases by a factor 5/3. Hence if the hydrocyclone operates satisfactorily when the feed concentration is 5% by weight, then the underflow needs to be increased by a factor of about 1.6 to remove the increased number of particles entrained in the boundary layer flow when the feed concentration is 15%. It should be noted that if this result is not complied with then the underflow aperture may well become blocked since the particle concentration, and hence the effective viscosity of the flow, increases rapidly near the vertex of the hydrocyclone. The effect of increasing d/λ^* leaves the general character of Figs. 32, 33 and 34 unchanged. The only differences arise as a result of the movement of the cut-off point to larger values of R'.

4.4 Distribution of particle sizes in the feed

The ideas used in the previous section can be extended to construct a mathematical model which is able to cope with a distribution of particle sizes in the feed. The modifications to the theory are outlined below.

It is convenient to consider the feed as consisting of a discrete number of different particle sizes rather than a continuous distribution which would be a more realistic model of the physical situation. Such an assumption is expected to retain the essential features of the problem while greatly simplifying the analysis. It is again necessary to determine which particles enter the side wall boundary layer since, assuming the particle concentration is low in the main flow, only those particles within the boundary layer can affect the flow. Now for each particle size, denoted by d_i, $i=1,\ldots,m$ and m is the number of different particle sizes, there exists an equilibrium line and these lines cross the line of zero vertical velocity at different points denoted by M_i, $i=1,\ldots,m$. For each value of d_i it is possible to integrate along its path from M_i to the homogeneous region and hence the particles of each size that enter the boundary layer flow can be found.

The equation which is equivalent to eqn. (4.27) for a distribution of particle sizes can easily be shown to be

$$Q_s \xi = -\sum_{i=1}^{m} \xi_{oi} \psi_i \qquad (4.39)$$

where ξ_{oi} is the mass of particles of diameter d_i per unit volume of the homogeneous suspension, and ψ_i is the stream function, at the level where the suspension is regarded as homogeneous, associated with the particles of diameter d_i.

All the remaining analysis carries through as before when one particle size was considered and hence is not presented here.

Computations were again performed for operating conditions appropriate to Kelsall's experimental data. As an example, the discrete particle size distribution considered was d/λ^*=0.05, 0.10, 0.15, 0.20 and 0.25 microns and the feed concentration of each size is taken to be 3% by weight. Now in practice there may be a wide range of particle sizes in the feed in which case the method outlined here would be very laborious. Hence it would be useful if one particle diameter could be chosen to represent the size distribution. Because the particular diameter d as a parameter usually appears as a square, the representative particle diameter chosen is the root mean square value d defined as

$$\bar{d}^2 = \frac{\sum_{i=1}^{m} (d_i d_i C_{oi})}{\sum_{i=1}^{m} C_{oi}} \qquad (4.40)$$

and is equal to 10.8 microns for the size distribution given above.

Fig.35 shows the particle concentration within the boundary layer, plotted as a function of R', for d=10.8 microns and for the given size distribution. It can be seen that the difference between the results are not significantly large. Similarly the Q_s through the boundary layer arising from consideration of the R.M.S. value of d deviates little from that due to the particle distribution (see Fig.36). Similar results were obtained when the concentrations of each particle size were not assumed the same. Hence it probably seems reasonable to use the results obtained from the representative particle size when drawing conclusions on the operation of a hydrocyclone.

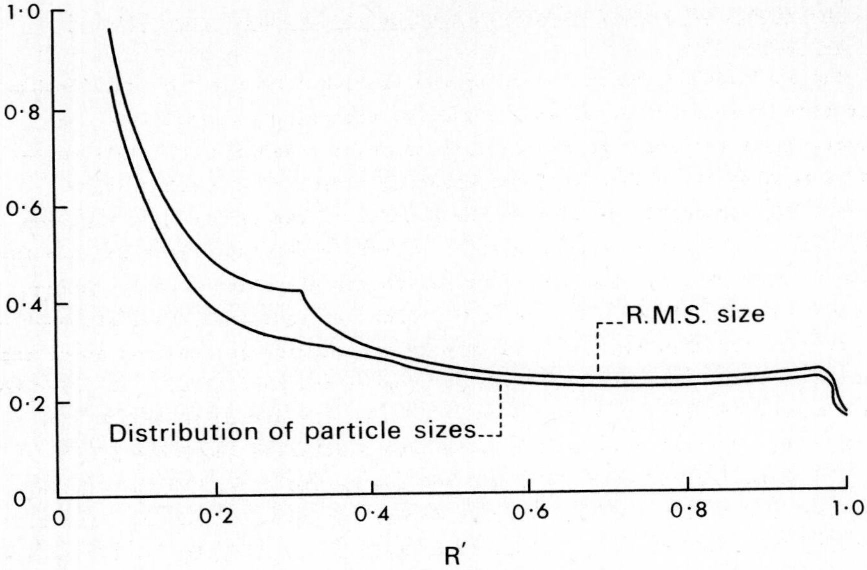

Fig.35 The particle concentration through the boundary layer as a function of R' for a distribution of particle sizes and the r.m.s. size

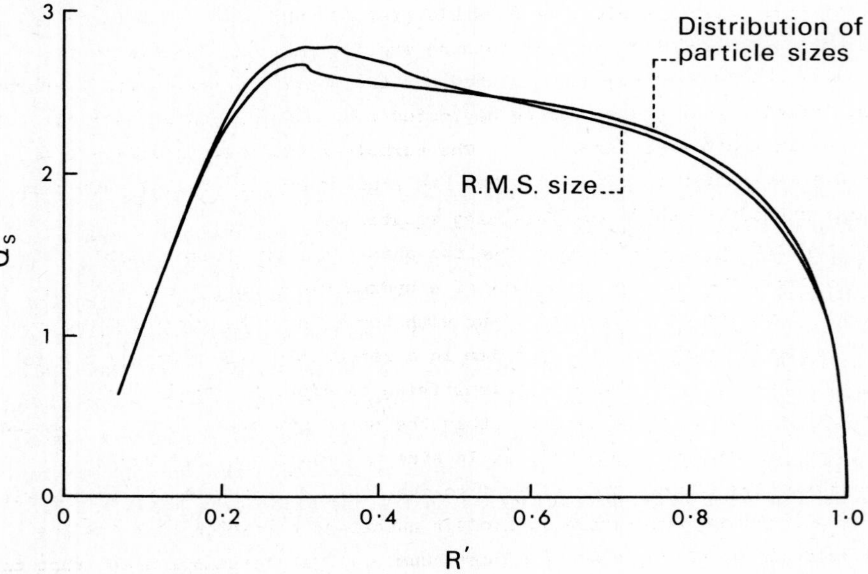

Fig.36 The meridional flux as a function of R' for a distribution of particle sizes and the r.m.s. size

4.5 The effect of particle concentration on the boundary layer flow — two phase flow

The analysis of the influence of the particles on the side wall boundary layer flow by means of a variable effective viscosity averaged over the boundary layer thickness makes interpretation of results a difficult matter. This is because there is a self compensating effect in the model which prevents a catastrophic failure in the boundary layer of the type expected physically in the blockage mechanism. As more particles are swept into the boundary layer with a consequent increase in the viscosity, the averaging process ensures that the boundary layer thickness increases, thus attenuating the trend towards blockage. Clearly more information is required about the detailed structure of the boundary layer containing particles. In order to gain a deeper understanding of the way in which the particles affect the boundary layer flow, it is necessary to have a detailed knowledge of the velocity components of both the fluid and the particles within the boundary layer and of the particle distribution across the boundary layer. Hence, following Laverack (ref.12) the boundary layer will be treated as two separate phases, namely a fluid phase and a particle phase, which interact with each other.

A considerable amount of progress has been made in recent years in the mechanics of two phase flow, the major impetus having come from the efforts of Soo (ref.39), Saffman (ref.40) and Marble (ref.41) who have established the equations of motion of a fluid containing small identical non-interacting particles. Chiu (ref.42) has examined the two phase boundary layer flow over a semi-infinite flat plate, but he neglected the conservation of particle momentum in the normal direction. The turbulent two phase boundary layer flow on a semi-infinite flat plate has bee studied by Soo (ref.43) who used integral methods to solve the governing equations.

The problem considered here is the two phase laminar incompressible boundary layer flow on the side wall of a hydrocyclone due, for simplicity, to a potential vortex which is coincident with the axis of the hydrocyclone.

In order to formulate the problem in a reasonably simple manner and to bring out the essential features, simplifying assumptions are made about the the motion of the solid particles within the boundary layer. It is supposed that the solid particles are uniform in size and shape and that their concentration by volume is very small so that they do not have collisions with each other, though there are a sufficient number of particles to allow the particle phase to be regarded as a continuum. It is further assumed that the flow field about each individual particle does not interact with the flow field about any other particle. Also, provided the mass concentration is at most unity and that the Reynolds number of the particles, based on the particle size and relative velocity, is small compared with unity, the net effect of the particles on the fluid can be described by Stokes' drag law.

Under these assumptions the boundary layer equations for axially symmetrical flow may be written

$$u \frac{\partial u}{\partial s} + w \frac{\partial u}{\partial n} - \frac{v^2}{s} = - \frac{1}{\rho} \frac{dp}{ds} + \frac{\mu}{\rho} \frac{\partial^2 u}{\partial n^2} + \tau\Theta(u_p - u) \quad (4.41)$$

$$u \frac{\partial}{\partial s} (sv) + w \frac{\partial}{\partial n} (sv) = \frac{\mu}{\rho} \frac{\partial^2}{\partial n^2}(sv) + \tau\Theta(sv_p - sv) \quad (4.42)$$

and the continuity equation is

$$\frac{\partial}{\partial s} (su) + \frac{\partial}{\partial n} (sw) = 0 \quad (4.43)$$

where

$$\Theta = \xi/\rho \quad (4.44)$$

and

$$\tau = 18\mu/d \, (\rho_s - \rho)^2 \quad (4.45)$$

The momentum equations for the particles are

$$u_p \frac{\partial u_p}{\partial s} + w_p \frac{\partial u_p}{\partial n} - \frac{v_p^2}{s} = - \tau \, (u_p - u) \quad (4.46)$$

$$u_p \frac{\partial}{\partial s} (sv_p) + w_p \frac{\partial}{\partial n} (sv_p) = - \tau \, (sv_p - sv) \quad (4.47)$$

$$u_p \frac{\partial w_p}{\partial s} + w_p \frac{\partial w_p}{\partial n} - \frac{v_p^2}{s} \cot \alpha^* = - \tau \, (w_p - w) \quad (4.48)$$

and the continuity equation for the particle phase is

$$\frac{\partial}{\partial s} (su_p\Theta) + \frac{\partial}{\partial n} (sw_p\Theta) = 0 \quad (4.49)$$

Here (u_p, v_p, w_p) are the velocity components of the particles in the (s, λ, n) directions respectively.

The boundary conditions to be satisfied by eqns. (4.41), (4.42) and (4.43) are

$$u = v = w = 0, \quad \text{on } n=0 \quad (4.50a)$$

and at the outer edge of the boundary layer

$$u \to U(s), \qquad v \to V(s), \tag{4.50b}$$

and the boundary conditions on the particle phase at the outer edge of the boundary layer are

$$\left.\begin{array}{ll} u_p \to U_p(s), & v_p \to V_p(s), \\ w_p \to W_p(s), & \theta \to \theta_0(s) \end{array}\right\} \tag{4.51}$$

where U_p, V_p, W_p and θ_0 are to be prescribed later.

It is convenient to non-dimensionalise eqns. (4.41) – (4.49) as before. The additional non-dimensional variables are defined as follows:

$$u_p' = u_p/U_i; \qquad v_p' = v_p/V_i; \qquad w_p' = \frac{18\nu s_0}{d^2(\rho_s/\rho - 1)\,V_i^2}\, w_p \tag{4.52}$$

Since the centrifugal force drives the particles through the boundary layer w_p has been non-dimensionlised with respect to a typical drift velocity.

Substituting the non-dimensional variables into the momentum equations and continuity equation for the fluid yields

$$u'\frac{\partial u'}{\partial s'} + w'\frac{\partial u'}{\partial n'} - K^2\frac{v'^2}{s'} = -\frac{dp'}{ds'} + \frac{\partial^2 u'}{\partial n'^2} + \frac{\theta}{\sigma Re}K^2(u_p' - u') \tag{4.53}$$

$$u'\frac{\partial}{\partial s'}(s'v') + w'\frac{\partial}{\partial n'}(s'v') = \frac{\partial^2}{\partial n'^2}(s'v') + K^2\frac{\theta}{\sigma Re}(s'v'_p - s'v') \tag{4.54}$$

Further, the boundary conditions given by equation (4.50) become

$$\left.\begin{array}{lll} u' = v' = w' = 0 & \text{on } n' = 0 & \\ u' \to -1/s'^{1/2}, & v' \to 1/s' & \text{as } n' \to \infty \end{array}\right\} \tag{4.55}$$

The momentum equations and the continuity equation for the particles become

$$u_p'\frac{\partial u_p'}{\partial s'} + \sigma Re^{3/2} w_p'\frac{\partial u_p'}{\partial n'} - K^2\frac{v_p'^2}{s'} = -K^2\frac{(u_p' - u')}{\sigma Re} \tag{4.56}$$

$$u'_p \frac{\partial}{\partial s'}(s'v'_p) + \sigma Re^{5/2} w'_p \frac{\partial}{\partial n'}(s'v'_p) = - K^2 \frac{(s'v'_p - s'v')}{\sigma Re} \tag{4.57}$$

$$\frac{\sigma Re}{K^2} u'_p \frac{\partial w'_p}{\partial s'} + \frac{\sigma^2 Re^{5/2}}{K^2} w'_p \frac{\partial w'_p}{\partial n'} + \frac{v'^2_p}{s'} \cot \alpha^* = - w'_p + \frac{w'}{\sigma Re^{3/2}} \tag{4.58}$$

and

$$\frac{\partial}{\partial s'}(s'u'_p\Theta) + \sigma Re^{3/2} \frac{\partial}{\partial n'}(s'w'_p\Theta) = 0 \tag{4.59}$$

The boundary conditions on u'_p, v'_p, w'_p and Θ are still unspecified at this stage.

In the above equations the Reynolds number is defined as

$$Re = \frac{\rho U_i s_o}{\mu} \tag{4.60}$$

and the non-dimensional parameter σ is

$$\sigma = \frac{K^2}{18} (\rho_s/\rho - 1)(d/s_o)^2 \tag{4.61}$$

The different ways in which a small particle can make a small perturbation to the boundary layer flow depends on their size. In particular, it can be deduced from the above equations that the various size ranges which need to be considered are

(1) $d^2/s_o^{\,2} = O(Re^{-11/8})$,

(2) $d^2/s_o^{\,2} = O(Re^{-3/2})$,

(3) $d^2/s_o^{\,2} \ll O(Re^{-3/2})$,

It is worth noting that for operating conditions appropriate to Kelsall's hydrocyclone (ref.3) the middle range represents particles of size around d_{50}. For particles in size range (3), the particles move with the fluid and hence the problem reduces to solving the particle free boundary layer equations except that the term $\partial^2/\partial n^2$ in the momentum equations is replaced by $(1 + \Theta)^{-1} \partial^2/\partial n^2$. Results for this case have been discussed earlier in section 3.3 and need not be discussed further.

For larger particles in size range (1), the particle motion is dominated by the centrifugal acceleration and the drag force normal to the wall has a negligible effect. The particle momentum equations (4.56) - (4.58) then reduce to

$$\left.\begin{aligned} u_p &= u' + O(Re^{-1/4}) \\ v_p' &= v' + O(Re^{-1/4}) \\ w_p' &= -\frac{v'^2}{s'}\cot\alpha^* + O(Re^{-1/8}) \end{aligned}\right\} \quad (4.62)$$

Because the full equations of motion have been reduced to the boundary layer equations, a boundary condition at large distances from the wall has had to be relaxed so $\Theta_o(s)$ cannot be chosen arbitrarily. In fact for this case $\Theta_o(s)$ must be taken as

$$\Theta_o(s') = \Theta_o(1) \,/\, s'^{1/2} \quad (4.63)$$

However, this variation of $\Theta_o(s')$ bears some resemblance to the practical situation. Thus the equations to be solved in this case are

$$(1+\Theta)u' \frac{\partial u'}{\partial s'} + (w'+\Delta_1 Re^{1/8}\Theta w_p') \frac{\partial u'}{\partial n'} - (1+\Theta)K^2 \frac{v'^2}{s'}$$

$$= -(1+\Theta_o)\left(\frac{1}{2s'^2} + \frac{K^2}{s'^3}\right) + \frac{\partial^2 u'}{\partial n'^2} \quad (4.64)$$

$$(1+\Theta)u' \frac{\partial}{\partial s'}(s'v') + (w'+\Delta_1 Re^{1/8}\Theta w_p') \frac{\partial}{\partial n'}(s'v')$$

$$= \frac{\partial^2}{\partial n'^2}(s'v') \quad (4.65)$$

$$\frac{\partial}{\partial s'}(s'u'\Theta) + \Delta_1 Re^{1/8} \frac{\partial}{\partial n'}(s'w_p'\Theta) = 0 \quad (4.66)$$

together with the eqns. (4.43) and (4.62) subject to the boundary conditions

$$\left.\begin{aligned} &u' = v' = w' = 0 \quad \text{on } n' = 0 \\ &u' \to -1/s'^{1/2}, \quad v' \to 1/s', \quad \Theta \to \Theta_o(1)/s'^{1/2} \quad \text{as } n' \to \infty \end{aligned}\right\} \quad (4.67)$$

where

$$\Delta_1 = \sigma\, Re^{11/8} \quad (4.68)$$

The method of solution is similar to that outlined in section (3.3).

The somewhat smaller particles in the size range (2) feel the centrifugal acceleration and experience a drag force exerted by the fluid. Thus, the particle momentum eqns. (4.56) - (4.58) become

$$\left.\begin{aligned} u'_p &= u' + O(Re^{-1/2}) \\ v'_p &= v' + O(Re^{-1/2}) \\ w'_p &= \frac{w'}{\Delta_2} - \frac{v'^2}{s'}\cot\alpha^* + O(Re^{-1/4}) \end{aligned}\right\} \tag{4.69}$$

where

$$\Delta_2 = \sigma\, Re^{3/2} \tag{4.70}$$

As explained in the previous case, for consistency, the choice of $\Theta_o(s')$ is restricted and must be taken as a constant. This restriction should not be regarded as serious as the present analysis is aimed at gaining some insight into the general behaviour of particles within the boundary layer. It follows that eqns. (4.53), (4.54), (4.59) and (4.69) become

$$(1+\Theta)u'\frac{\partial u'}{\partial s'} + (w'+\Delta_2\Theta w'_p)\frac{\partial u'}{\partial n'} - (1+\Theta)K^2\frac{v'^2}{s'}$$

$$= -(1+\Theta_o)\left(\frac{1}{2s'^2} + \frac{K^2}{s'^3}\right) + \frac{\partial^2 u'}{\partial n'^2} \tag{4.71}$$

$$(1+\Theta)u'\frac{\partial}{\partial s'}(s'v') + (w'+\Delta_2\Theta w'_p)\frac{\partial}{\partial n'}(s'v')$$

$$= \frac{\partial^2}{\partial n'^2}(s'v') \tag{4.72}$$

$$\frac{\partial}{\partial s'}(s'u'\Theta) + \Delta_2\frac{\partial}{\partial n'}(s'w'_p\,\Theta) = 0 \tag{4.73}$$

$$w'_p = \frac{w'}{\Delta_2} - \frac{v'^2}{s'}\cot\alpha^* \tag{4.74}$$

respectively. These together with the fluid continuity eqn. (4.43) need to be solved subject to the boundary conditions

$$\left.\begin{array}{l} u' = v' = w' = 0 \quad \text{on } n' = 0 \\ u' \to -1/s'^{1/2}, \quad v' \to 1/s', \quad \Theta \to \Theta_o(1) \quad \text{as } n' \to \infty \end{array}\right\} \qquad (4.75)$$

Again, the method of solution is similar to that outlined in section (3.3).

The results obtained when the particles are in size range (1) are shown in Figs. 37, 38, 39. In these figures s'u', s'v' and Θ are plotted against n' at three stations along the hydrocyclone wall, for various values of the parameter Δ_1, with $\Theta_o(1)=0.05$. Concentrating first on Figs. 37a, 37b and 37c, it is clear that an interchange of momentum between fluid and particles exists, and the particles tend to drag on the fluid as it is accelerated within the overshoot region. Obviously the "larger" particles sap more momentum from the fluid than do the smaller particles and this is clearly illustrated in Fig.37 if Δ_1 is taken to be a function of d only. Once the particles have passed through the overshoot region then the roles are reversed so that the particles now accelerate the fluid since they are moving into a region of decelerating flow. The larger particles carry more momentum into this region than do the smaller particles and hence they have a greater influence on the flow.

Coming now to the spin component of velocity, shown in Fig.38, when there are no particles within the flow the fluid simply decelerates from its mainstream value to zero at the wall. Since the particles carry momentum and move with the fluid at the outer edge of the boundary layer then they impart some of their momentum to the fluid as they cross the boundary layer and therefore accelerate the flow in the azimuthal direction.

As a consequence of the assumption regarding the particles in this size range Θ becomes singular at the wall. Also since the centrifugal force acting on the particles increases as Δ_1 increases then as the solution progresses down the hydrocyclone wall the larger particles tend to be concentrated into a narrower region near the wall than the smaller particles (see Fig.39c).

The general effects of increasing the feed concentration $\Theta_o(1)$ on the velocity profiles and the particle distributions across the boundary layer are similar to the effects of increasing particle size shown in Figs.37,38 and 39. This is because increasing the number of particles within the boundary layer has the effect of further decreasing the overshoot velocity since more of the fluid momentum is needed to maintain the flow of the particles. Conversely, increasing the particle concentration causes an increase in the azimuthal velocity component because of the extra momentum supplied by the particles.

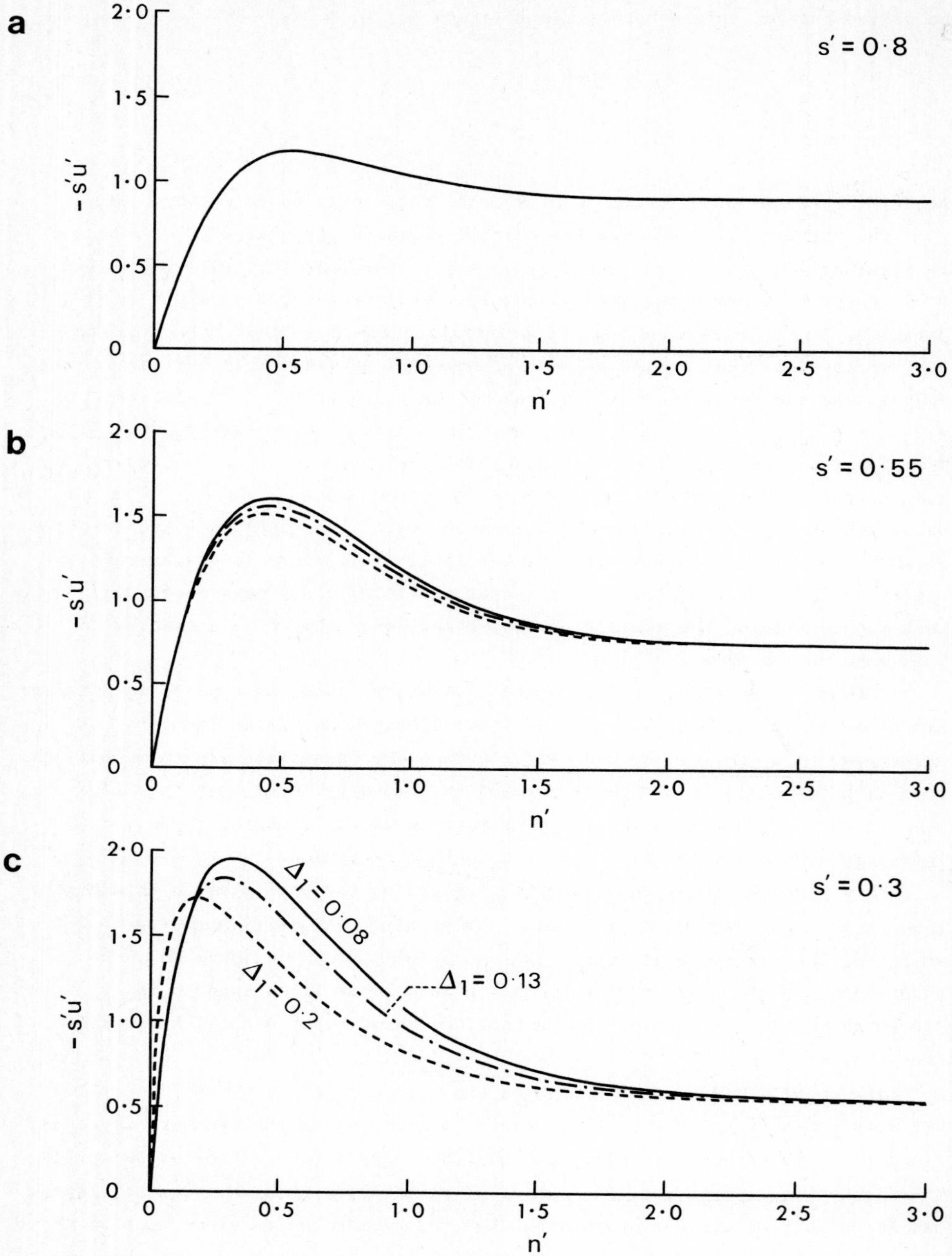

Fig.37 Plot of -s'u' against n' for various values of Δ_1 with $\Theta_0(1)=0.05$.

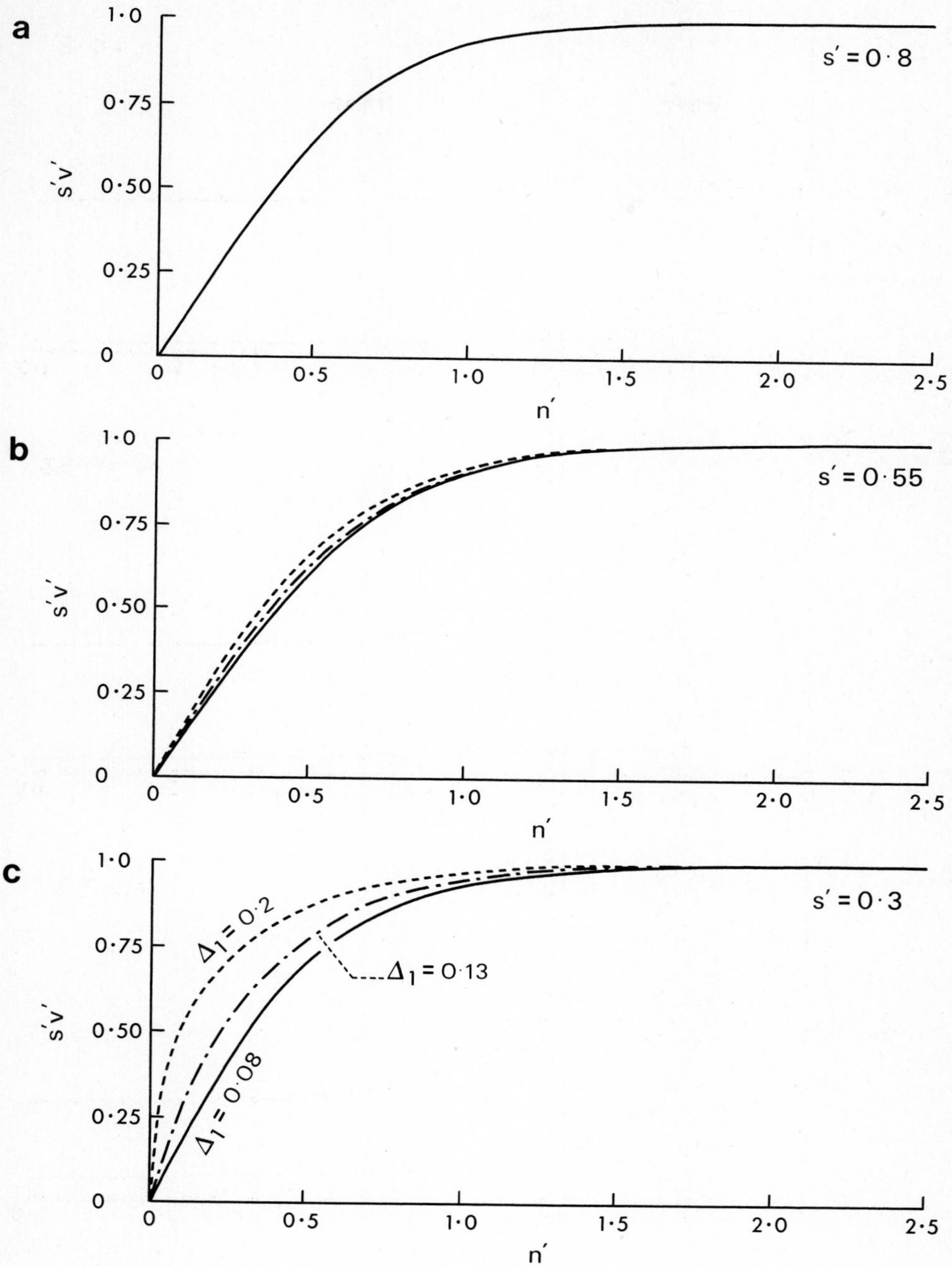

Fig.38 Plot of s'v' against n' for various values of Δ_1 with $\Theta_0(1)=0.05$.

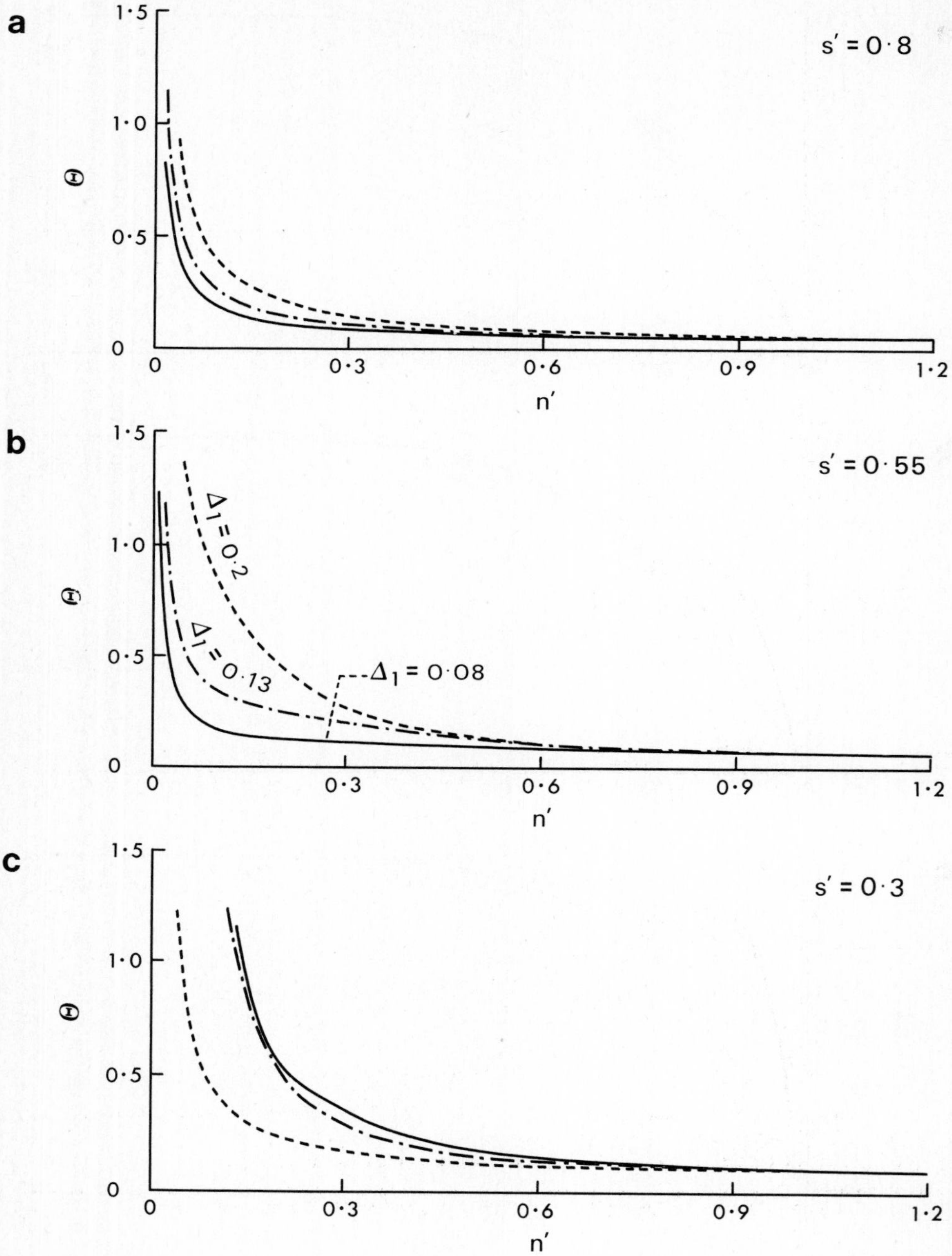

Fig.39 Plot of Θ against n' for various values of Δ_1 with $\Theta_0(1)=0.05$.

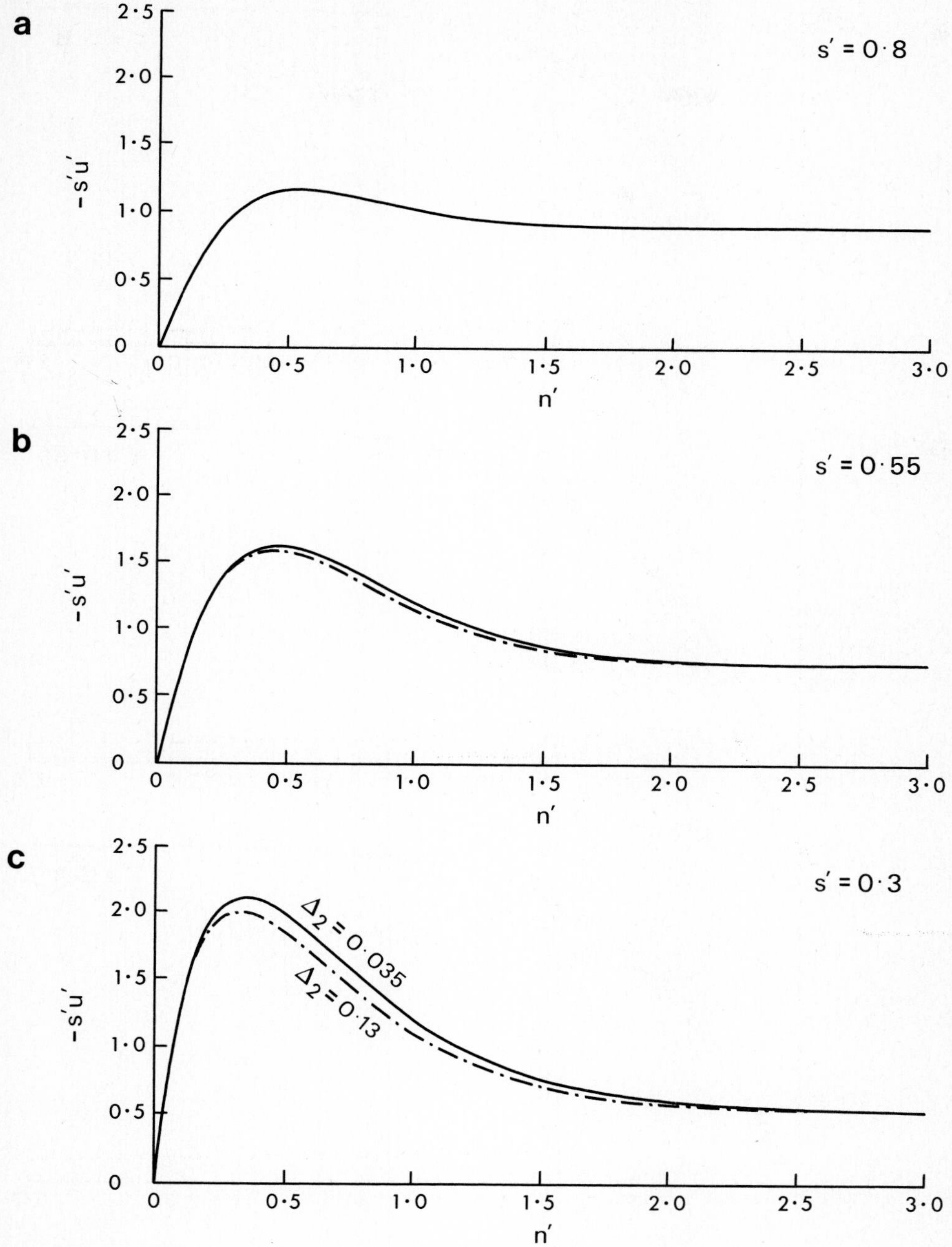

Fig.40 Plot of $-s'u'$ against n' for various values of Δ_2 with $\Theta_0(1)=0.05$.

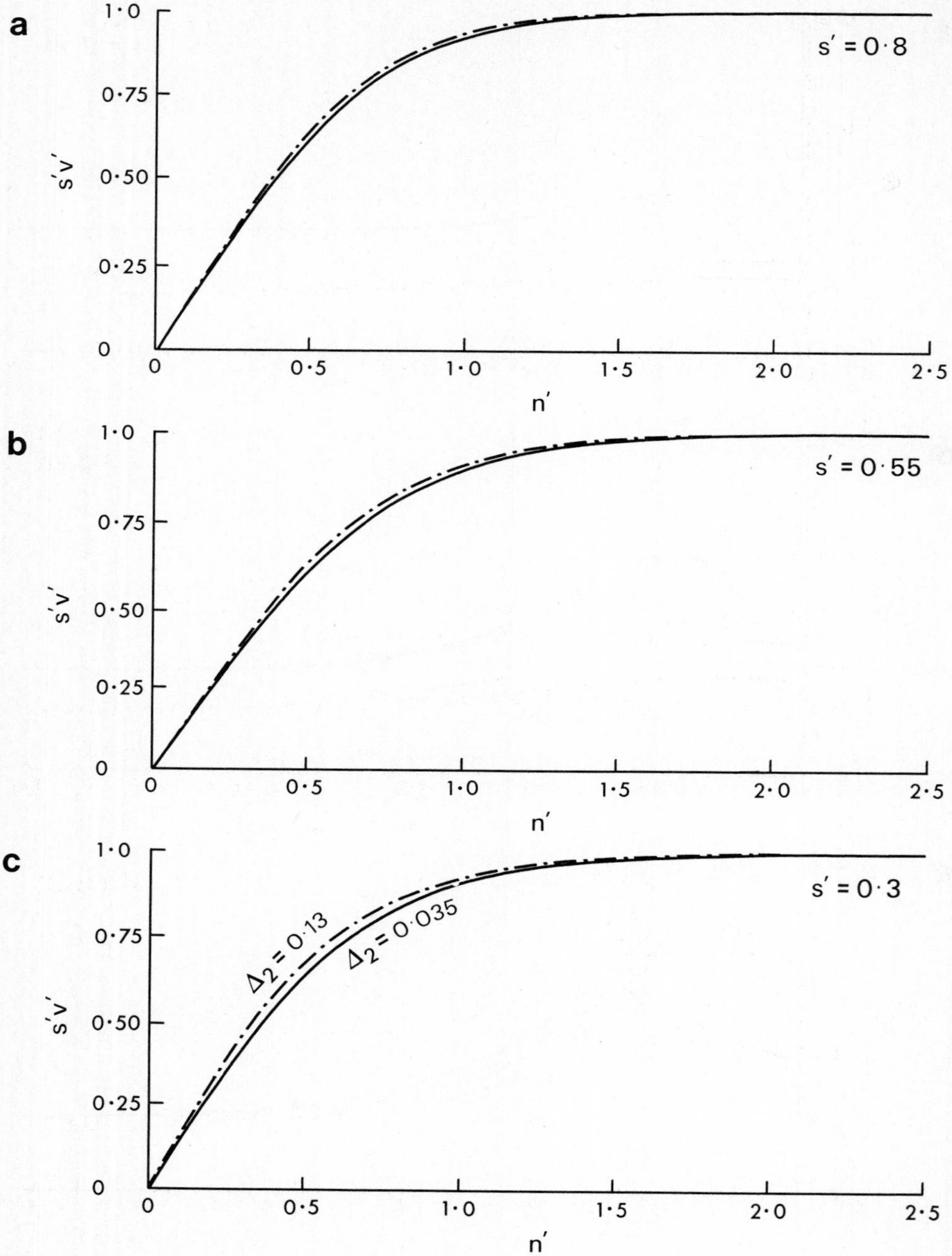

Fig.41 Plot of s'v' against n' for various values of Δ_2 with $\Theta_0(1)=0.05$.

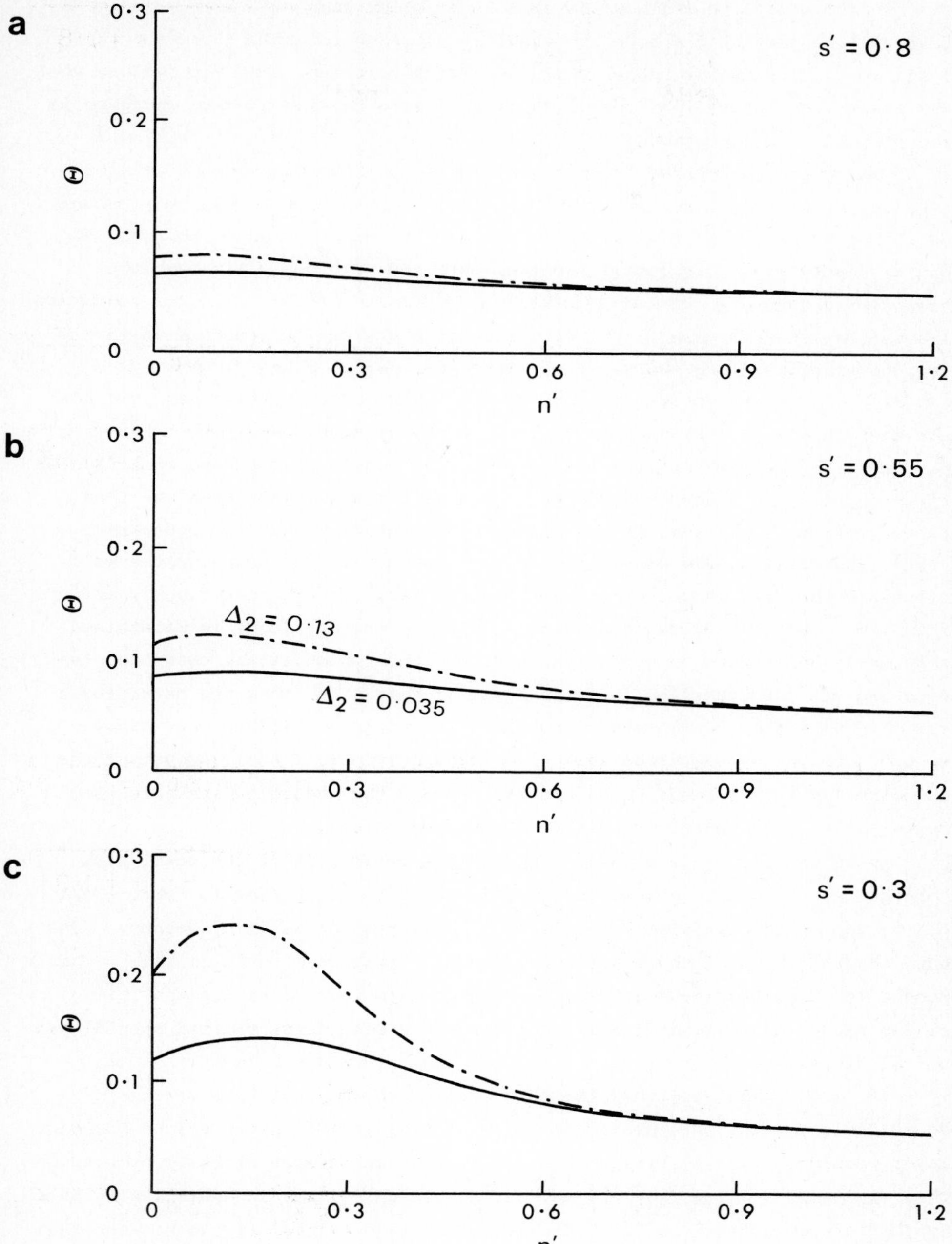

Fig.42 Plot of Θ against n' for various values of Δ_2 with $\Theta_0(1)=0.05$.

These results in which w' is of a lower order than w'_p clearly indicate that a large part of the boundary layer is not significantly altered by the particles. The narrow region which is affected is near the hydrocyclone wall where $s'v'$, and hence w'_p, reduce rapidly to zero causing a sharp increase in the particle concentration.

Coming now to the results obtained for size range (2), the effects of varying the parameter Δ_2 on su, sv and Θ are illustrated in Figs. 40, 41 and 42. The variation of $s'u'$ with n', and of $s'v'$ with n', shown in Figs. 40 and 41 respectively, is similar to that obtained previously although the effect of changing Δ_2 is not as appreciable due to the fact that the particles are smaller in this case. The variation of Θ across the boundary layer, at various stations along the wall, for different values of Δ_2 is shown in Fig.42. Again, the particle concentration is seen to increase near the wall although it does not become singular as in the previous case.

The results presented can be used to draw conclusions on the operation of a hydrocyclone. Since the present theory is not applicable when the feed concentration is low then the particle concentration within the boundary layer, and particularly within the narrow region near the hydrocyclone wall discussed earlier, is not expected to become large enough to seriously affect the flow. Hence provided the swirling flow in the main body is maintained throughout the entire length of the hydrocyclone, which is the case when the discharge at the underflow is vortex-like or rope-like, then the centrifugal force is expected to constrain the particles to remain within the narrow region near the hydrocyclone wall. Consequently only a small amount of the boundary layer flow needs to be removed through the underflow aperture to capture all of the particles within the boundary layer.

As an example it is expected that as little as 1.5% of the total flux through the hydrocyclone needs to be removed at the underflow to capture all of the separated particles when the feed concentration is 1% by weight. The mass concentration of the underflow product in this case would be 75%. This result is of particular significance when the aim is to minimise the contamination of the overflow product since the feed concentration should then be relatively low.

It is not suggested that the full scale numerical solution of the equations of motion is a practical means of tackling the hydrocyclone boundary layer problem. Its justification is that it enables some light to be shed on this extremely complex flow and may offer some guidelines in establishing more approximate methods. In fact, from the results obtained, it can be seen that the boundary layer consists of essentially two layers, an inner highly viscous slurry and an outer layer of relatively low particle concentration. This suggests that by making plausible assumptions about the particle and fluid velocity distributions across each sub-layer the well tried Pohlhausen technique could be used to solve the momentum integral equations. This work is in progress (ref.22) and the initial results look promising.

NOMENCLATURE

Latin

a	outer radius of vortex finder
c	concentration by weight of the solid particles in the feed
c_m	maximum concentration
d	equivalent spherical Stokes' diameter of particle
$\bar{d}$	root mean square value of d
d_{50}	diameter of particles of which 50% go to the overflow and 50% to the underflow
D	typical linear dimension of inlet
$\underline{D}$	Stokes' drag = $3\pi\mu\underline{U}d$
$D_h(d)$	modified distribution of particles due to leakage of particles
$D_o(d)$	distribution of particles in the fluid that enters hydrocyclone
E	overall efficiency of hydrocyclone
f_0, f_1, f_2	velocity profiles
g	spin velocity profile
H	$=(1/R_c)(1/(-U_o\nu s_o))^{1/2}\ Qm^{5/2}/2\pi(m-1)$
K	$= -V_o/U_o$
ℓ	distance of end of vortex finder from vertex of cone
L	non-dimensional parameter = $\rho BR_o{}^{5/2}/\mu_t$
m	R_c/R_o
$\underline{n}$	unit vector along inward normal to wall of hydrocyclone
p	pressure
p^*	modified pressure
$\underline{q}$	fluid velocity
q_r, q_θ, q_λ	velocity components in (r,θ,λ) coordinates
q_R, q_λ, q_z	velocity components in (R,λ,z) coordinates
Q	flux of fluid through hydrocyclone
Q_s	flux of fluid through side wall boundary layer
Q_L	flux of fluid through top wall boundary layer
r	spherical polar distance from apex of cone
R	cylindrical radial coordinate
R_c	radius of hydrocyclone

R_e	Reynolds number = $\rho U_i s_o / \mu$
R_f	ratio of underflow to feed flow rate
R_o	radius of hydrocyclone at level of vortex finder
R_p	particle Reynolds number = $\lvert\underline{U}\rvert d/\nu$
s	distance along cone from apex of hydrocyclone
u,v,w	velocity components in (s,λ,n) directions
U,V	velocities at edge of boundary layer in s and λ directions respectively
$\underline{U}$	fluid velocity relative to a particle
U_d	drift speed of solid particles in R direction
y	mesh scaling in boundary layer
z	distance along axis of hydrocyclone from apex

<u>Greek</u>

α	loss factor
α^*	semi angle of cone
$\gamma(n,x)$	Incomplete Gamma Function of order n and argument x
Γ	Complete Gamma Function
δ	boundary layer thickness
Δ_1	$= \sigma\ Re^{11/8}$
Δ_2	$= \sigma\ Re^{3/2}$
ϵ	$= 1 - R'$
η	a similarity variable in Pohlhausen method
ξ	mass of particles per unit volume of the suspension
ζ	a similarity variable in numerical solution
θ	cylindrical polar angle
Θ	$=\xi/\rho$
λ	azimuthal angle in spherical co-ordinates
λ^*	particle size parameter
λ_1, λ_2	parameters in the boundary layer velocity profiles
Λ	a mixing length
μ	coefficient of viscosity of fluid
ν	kinematic viscosity of fluid
ν_s	kinematic viscosity of slurry
ρ	density of fluid
ρ_s	density of solid particles
σ	$= \frac{K^2}{18}\left(\frac{\rho_s}{\rho} - 1\right)\left\{\frac{d}{s_o}\right\}^2$
σ^*	$= \pi/2$ or α^*

τ	$= 18\mu/\{d^2/(\rho_s - \rho)\}$
φ	velocity potential
χ	potential energy per unit mass
ψ	stream function
$\underline{\omega}$	vorticity of fluid
Constants	A, B, C_1, C_2, E, F, h, k, K, M, S, β, γ, σ_1, σ_2
Integrals	D_1, D_2, D_3, I_{ij}, I_i, J_i
Suffices	
i	inlet conditions
o	conditions at the level of bottom of vortex finder
p	particle properties

REFERENCES

1. D. Bradley, The Hydrocyclone, 1st. Edn., Pergammon, Oxford, 1965.
2. F. J. Fontein and C. Dijksman, Inst. Min. and Metall. Symp. Min. Dressing, London, 1953, 229–247.
3. D. F. Kelsall, Trans. Inst. Chem. Engrs., 30 (1952) 87–103.
4. K. Rietema, Chem. Engng. Sci., 15 (1961) 298–319.
5. M. I. G. Bloor and D. B. Ingham, Proc. 1st. European Conference on Mixing and Centrifugal Separation, Cambridge, England, 1974, E6-95–E6-106.
6. M. I. G. Bloor and D. B. Ingham, Trans. Inst. Chem. Engnrs., 51 (1973) 36–41.
7. M. I. G. Bloor and D. B. Ingham, Trans. Inst. Chem. Engnrs., 53 (1975) 1–6.
8. M. I. G. Bloor and D. B. Ingham, J. Applied Maths. Phys., 28 (1977) 289–299.
9. M. I. G. Bloor and D. B. Ingham, Trans. Inst. Chem. Engnrs., 54 (1976) 275–280.
10. M. I. G. Bloor and D. B. Ingham, Trans. Inst. Chem. Engnrs., 53 (1975) 7–11.
11. S. D. Laverack, Trans. Inst. Chem. Engnrs., 58, (1980) 33–42.
12. S. D. Laverack, Ph.D. Thesis, University of Leeds, Leeds, England, 1980.
13. F. B. Hildebrand, Advanced Calculas for Applications, 5th. Edn., Prentice-Hall, Englewood Cliffs, New Jersey, 1965, 646pp.
14. H. Schlichting, Boundary Layer Theory, 4th. Edn., McGraw-Hill, London, 1960, 647pp.
15. B. Thwaites, Incompressible Aerodynamics, 1st. Edn., Oxford Univ. Press, Oxford, 1960, 636pp.
16. N. Rott and W. S. Lewellen, Prog. Aeronaut. Sci., 7 (1966) 111–144.
17. G. I. Taylor, Q. J. Mech. Appl. Math., 3 (1950) 129–139.
18. A. M. Binnie and D. P. Harris, Q. J. Mech. Appl. Math., 3 (1950) 89–106.
19. J. C. Cooke, J. Aeron. Sci., 19 (1952) 486–490.
20. J. C. Burggraf, K. Stewartson and R. J. Belcher, Phys. Fluids, 14 (1971) 1821–1833.
21. G. Wilks, J. Fluid Mech., 34 (1968) 575–563.
22. G. N. Gaunt, private communication.
23. W. R. Sears, J. Aeron. Sci., 15 (1948) 49–52.
24. M. I. G. Bloor and D. B. Ingham, Phys. Fluids, 20 (1977) 1228–1233.

25. J. H. Merkin, J. Fluid Mech., 35 (1969) 437-450.
26. J. Abrahamson, C. G. Martin and K. K. Wong, Trans. Inst. Chem. Engrs., 56 (1978) 168-177.
27. M. L. Rosenzweig, D. H. Ross and W. S. Lewellen, J. Aerospace Sci., 29 (1962) 1142-1143.
28. T. von Karman, Z. Angew. Math. Mech., 1 (1921) 233-252.
29. G. K. Batchelor,An Introduction to Fluid Mechanics, Cambridge University Press, Cambridge, 1967, 615pp.
30. S. K. Karanfilian and T. J. Kotas, Proc. R. Soc. London, A376 (1981) 525-544.
31. D. Bradley, Ind. Chemist, 34 (1958) 473-478.
32. D. Bradley and and D. J. Pulling, Trans. Inst. Chem. Engnrs., 37 (1959) 34-45.
33. E. O. Lilgé, Trans. Inst. Min. and Metall., 71 (1962) 285-337.
34. G. Tarjan, Acta Tech. Hung., 32 (1961) 357-388.
35. G. Huyet, J. Grassaud and P. Solety, Revue Ind. Minér., 53 (1971) 574-583.
36. D. A. Dahlstrom, Chem. Engng. Prog. Symp. Ser. No. 15, 50 (1954) 41-47.
37. A. Hundertmark, Zeits. fur Erzbergbau. und Metall., 18 (1965) 403-412.
38. R. Every, R. Hughes and L. Koch, Advances in Solid-Liquid Flow in Pipes and Applications, Ed. Zandi, Pergammon, Oxford, 1971, 165pp.
39. S. L. Soo, University of Illinois Project Report, ILL-3P, 1961.
40. P. G. Saffman, J. Fluid Mech., 13 (1962) 120-128.
41. F. E. Marble, Ann. Rev. Fluid Mech., 2 (1970) 397-446.
42. H. H. Chiu, Princeton University Report 620, 1962.
43. S. L. Soo, Fluid Dynamics of Multiphase Systems, Blaisdell, London, 1967.

HIGH-INTENSITY, HIGH-GRADIENT DIELECTROPHORETIC FILTRATION AND SEPARATION PROCESSES

I.J. LIN[1] and L. BENGUIGUI[2]
[1]Mineral Eng. Res. Center. Technion - IIT, Haifa, 32000, Israel.
[2]Solid State Institute, Technion - IIT, Haifa, 32000, Israel.

CONTENTS

CONTENTS (continued)

ABSTRACT

The paper describes the general principle, operating characteristics, and typical performance data of dielectrophoretic devices.

High-gradient electric separation (HGES) and dielectric filtration (DF) are new techniques based on use of the polarization force exerted by a non-uniform electric field (dielectrophoretic effect). Matrices of finely divided filamentary dielectric material and other types (glass fabric, titanates in the form of beads, balls or rods, perovskite-type ferroelectric bodies, ceramic wool, glass beads, pads, etc.) containing 50 to 95% void space are used. The system requires an electric cell capable of generating high-intensity fields in large volumes.

Dielectric separation is discussed under the following aspects: (i) intensity of the electric field and its gradient; (ii) size, density and dielectric constant of the particles; (iii) dielectric constant of the fluids serving as separation media; and (iv) flow regime in the separator system.

The enormous potential of this technology is still to be realized. The most promising areas of application for the unique processes are identified.

INTRODUCTION

Magnetic and electric separation techniques for mineral and other materials are known, and have been in routine use for many years. Prior to the advent of the high-gradient magnetic separation (HGMS) and high-gradient electric separation (HGES) processes, these unique methods were inapplicable to fine particles.

Potential applications of both the magnetic and electric separation techniques in beneficiation of fine-particle feeds were reviewed (ref.1-3). Special features of these techniques were considered and compared with a view to increased amenability of such feeds to separation in an electromagnetic field (see Table 1).

Numerous researchers have theoretically and experimentally studied collection of paramagnetic and ferromagnetic particles from a fluid, using the HGMS method. Recently, HGMS has been used to capture diamagnetic particles by adjusting the magnetic susceptibility of the fluid (ref.16-17). In addition, magnetic separation may be used to separate liquid mixtures in two cases: (i) when the continuous phase is paramagnetic and the discrete phase is diamagnetic, and (ii) when the continuous phase is diamagnetic and the discrete paramagnetic.

TABLE 1

Some new technologies in magnetic and electric separation

Magnetic		Electric	
Technique	Ref.	Technique	Ref.
Magnetohydrostatic levitation* (MHSL)**	4	Dielectric levitation* (DL)**	6-9
Wet high-intensity magnetic separation (WHIMS)	2	Wet high-intensity dielectric separation (WHIES)	1,3,10-11
High-gradient magnetic separation (HGMS)	2	High-gradient dielectric separation (HGES)	3, 11-13
Magnetic filtration (MF)	2	Dielectric filtration (DF)	3, 10-11
Stream splitter magnetic separators*	5	Stream splitter dielectric separator*	14-15

* Matrices free devices (open gradient)
**Provide an artificial gravity useful in "zero-g" conditions

Control of paramagnetic aqueous fluids is obtained by adding a paramagnetic salt to water (ref.18).

The advent of new technologies of magnetic separation in solid-solid and solid-liquid systems, has generated increasing interest in the behaviour of similar methods in dielectrophoretic separation and filtration for S-S, S-L, L-L, S-G and L-G systems. All eight methods, listed in Table 1, involve a non-uniform field, so that the particles are subjected to ponderomotive (gradient) forces.

Few of the mineral beneficiation techniques are applicable to separation of fine particulates. Certain mixtures can be separated by exploiting specific differences - the dielectrophoretic method, for example, uses the difference in dielectric constant as the main parameter. Tables 2 and 3 show values of relative permittivities (dielectric constants) of some liquids, minerals, ceramics, plastics and other materials. It is possible, by introduction of certain materials (e.g. via adsorption or seeding), to change the dielectric characteristics and thereby create the conditions needed for separating loose particles, which would otherwise be unfeasible. To give an idea about the order of the difference in polarizability of various particles we quote the typical static dielectric constants of barium titanate ($BaTiO_3$)>1500, sodium niobate ($NaNbO_3$) 650, and barium sulphate ($BaSO_4$) 11.

One must be very wary about drawing any conclusions about how a material will behave in dielectric separation from tabulated ε values. For example, minerals do not have fixed chemical composition. Small departures from exact stoichiometry can make dramatic differences in electric properties. Sometimes materials have associated with themselves trace amounts of metal impurities, the effect of which can completely mask the intrinsic ε of the material. The quickest and surest way of finding out how a material will behave in a dielectrophoretic separator is to run it through the separator.

Electrical properties of minerals and rocks are of interest not only for electrodynamic-, electrostatic-, and dielectrophoretic-separation techniques but also because fragmentation by electrical means may be one of the practical methods available for use in size-reduction process (ref. 19-20). Also, dielectric heating as a method of rock fragmentation of non conducting materials and as a process for preconditioning (non uniform and selective heating) of mixed granular solids is mainly based on the dielectric properties. Studies of the relationship between dielectric constant, dissipation factor and frequency, and temperature, have been published (ref. 21-23). These include studies of rocks and minerals, glasses and ceramics, plastics and fluids.

TABLE 2

Typical Values of the Relative Dielectric Constants of Some Common Materials*

Substance	ε	Substance	ε
Air & other gases	~1	Ordinary glass	5.4-9.9
Kerosene	2.0	Apatite	5.72-10.5
Transformer oil	2.23	Salt (halite), NaCl	5.9
Benzene	2.28	Beryl, $Al_2Be_3(SiO_3)_6$	6.08-7.02
Carbon tetrachloride	2.3	Tourmaline	6.3-7.1
Corn oil	3.1	Calcite, $CaCO_3$	6.36
Castor oil	4. 3-4, 7	Magnesite, $MgCO_3$	6.62-7.36
Chloro-benzene	5. 62-6,0	Gypsum, $CaSO_4.2H_2O$	6.83
Ethanol	24.3	Dolomite, $CaMg(CO_3)_2$	6.8 - 8.0
O-Nitrotoluene	27.4	Siderite, $FeCO_3$	7 - 7.7
Methanol	33.5	Fluorite, CaF_2	7.11
Propylene carbonate	69.0	Barite, $BaSO_4$	8.24 - 10
Water	78.5	Spinel, $MgAl_2O_4$	8.4
Formamide	109	Alumina, Al_2O_3	8.5
Hydrogen cyanide	114.9	Corning 8870 (glass)	9.5
		Ruby	11.2 - 13.2
Teflon	2.0	Monazite,(Ce,La)PO_4	11.5 - 18.2
Styrene	2.4-4.75	Huebnerite, $MnWO_4$	12.38-13.7
Quartz, SiO_2	3.5-3.8	Zircon, $ZrSiO_4$	12.5
Mica	3.0-5.97	Wolframite, (Mn,Fe)WO_4	18
Talc	3.9 - 5.8	Cassiterite, SnO_2	27
Sulphur, S	4.0	Graphite	>81
Cellulose acetate	4.0 - 5.5	Rutile, TiO_2	86 - 170
Diamond, C	4.6 - 5.5	Lithium sulfate monohydrate	350
Sylvite, KCl	4.8 - 5.7	Barium Titanate**	>1500
Microcline, K[(Al,Si)$_3O_8$]	5. 38-5.6		

*See also Table 6

** Special ferroelectric properties. The temperature of the transition between ferroelectric and paraelectric phases (Curie Point) is 121°C. $BaTiO_3$ has a dielectric constant as high as 10 000 at elevated temperatures.

TABLE 3

Values of the Dielectric Constants of Some Plastics

Material	Sp.Grav. g/cc	Dielectric Constant 60 HZ	1,0 KHZ	1,0 MHZ	1,0 GHZ
Thermoplastic Molding Materials:					
Polyacetal	1.42	3.8		3.8	3.8
Acrylic	1.20	4.0		3.5	3.2
Nylon	1.14	5.5		4.9	4.7
Polycarbonate	1.2	3.2		3.0	3.0
Polyester	1.38	3.3		3.4	3.2
Polyimide	1.43	3.43		3.42	3.42
Polypropylene	0.91	2.6		2.6	2.6
Polytetra fluoro-ethylene	4.5	2.1		2.1	2.1
Polyvinylchloride	2.3	9.0		8.0	5.0
Thermoset Molding Materials:					
Diallyphthalate (with glass fiber)	1.78	4.3		4.4	4.5
(with mineral)	1.68	5.2		5.3	4.0
Epoxy (with glass fiber)	2.0	5.0		5.0	5.0
(with mineral)	2.0	5.0		5.0	5.0
Melamine (α-cellulose)	1.52	9.5		9.2	8.4
Polyester (with glass fiber)	2.3	7.3		4.7	6.4
(with mineral)	2.3	7.5		6.2	5.5
Silicon (with glass fiber)	2.0	5.2		5.0	4.7
(with mineral)	2.82	3.6		3.2	3.2
Urea Formaldehyde (α-cellulose)	1.52	9.5		7.5	6.8
Fluorinated Plastics:					
Polytetrafluoroethylene (PTFE)	2.13-2.20		2.0		
Poly(vinylidene fluoride)-PVF	1.78		8.5		
Polyolefin Resins:					
Polyethylene	0.91-0.96		2.3	2.2	2.2
Polypropylene copolymer	0.905		2.25		

In this paper a new separation method is presented, whereby mixtures of uncharged particulate and granular solids with widely differing dielectric constants of their constituents, can be efficiently separated by means of non-uniform electric fields (of a strength of a few kilovolts per cm). Insulating as well as conducting materials can be separated by both HGES and DF. Possible applications of high-gradient electric separation and dielectric filtration techniques are reviewed with regard to the beneficiation of fine-particle feeds, due attention being paid to the forces and media available.

In this paper we shall also discuss the theoretical basis of the dielectrophoretic technique and the published experimental work. It will be demonstrated that further theoretical and experimental work is required before a thorough understanding of the technique is available.

BASIC CONCEPTS

The amenability of ores, mixtures, suspensions and airborne particles to processes of separation, upgrading and cleaning, by means of electromagnetic fields depends on the following variables (given here for dielectric separation) which may be expected to play an important role:

(a) Characteristics of the Material Involved
- size and shape of particles
- dielectric constant of particles relative to that of the liquid medium
- electric conductivity of particles
- degree of liberation
- specific gravity
- chemical stability
- conductivity conditions and viscosity of the suspension
- fluid temperature

(b) Type of Forces Involved
- dielectric force
- drag force
- gravitational force
- London/van der Waals force
- double layer force
- inertial force

(c) Characteristics of the Separation System
- electric field data (frequency, voltage, gradient, intensity)
- relative direction of electric gradient
- profile and space geometry of electrodes
- separation cell geometry and working space
- matrix design (size, composition, porosity, thickness)

(d) Operational Parameters
- range of electric field
- type of liquid medium
- particle size range and concentration (solid content)
- static or batch separation process
- continuous separation
- suspension flow rate (residence time of particles)

The types of forces provided by an electromagnetic field, and their technological potential for application, are given in the sequel. Let us review briefly the forces involved in interaction of the electromagnetic field and the particles. The resultant electromagnetic force F_{em} is given by:

$$F_{em} = (M\cdot\nabla)H + (p\cdot\nabla)E + J \times B + q\cdot E \qquad (1)$$

where ∇ is the gradient operator, q the particle charge, M, H, B the magnetization, magnetizing force, and magnetic field intensity, respectively, J the current density and p, E, the polarization and electric field intensity.

Resorting to vector identities, equation (1) can be put in the following useful form.

$$F_{em} = \frac{1}{2}\nabla(M\cdot H) - M\times\nabla\times J + \frac{1}{2}\nabla(p\cdot E) + p\times\frac{\partial B}{\partial t} + J\times B + q\cdot E \qquad (2)$$

where t denotes time.

In equation (2) the electromagnetic force comprises six components, each of which with its own physical significance. Assigning to each a numerical index i = 1,2,....,6 in the above order, the respective forces per unit volume are given by:

$$F_1 = \frac{3}{8\pi}\,\frac{\mu_f(\mu_p-\mu_f)}{\mu_p + 2\mu_f}\quad\nabla(H^2) \qquad (3)$$

$$F_2 = \frac{3}{4\pi}\,\frac{\mu_f(\mu_p-\mu_f)}{\mu_p + 2\mu_f}\quad H\times\nabla\times J \qquad (4)$$

$$F_3 = \frac{3}{8\pi}\,\frac{\varepsilon_f(\varepsilon_p-\varepsilon_f)}{\varepsilon_p + 2\varepsilon_f}\quad\nabla(E^2) \qquad (5)$$

$$F_4 = \frac{3}{4\pi}\,\frac{\varepsilon_f(\varepsilon_p-\varepsilon_f)}{\varepsilon_p + 2\varepsilon_f}\quad E\times\frac{\partial B}{\partial t} \qquad (6)$$

$$F_5 = J\times B \qquad (7)$$

$$F_6 = q\cdot E \qquad (8)$$

where $\mu_f = 1 + 4\pi\chi_f$; $\mu_p = 1 + 4\pi\chi_p$

$$\varepsilon_f = 1 + 4\pi\chi_f^e \ ; \quad \varepsilon_p = 1 + 4\pi\chi_p^e$$

μ, χ and ε, χ^e denoting permeability, magnetic susceptibility - and dielectric constant, electric susceptibility, respectively. Subscripts f, p refer to the fluid (f) and particle (p), respectively.

In the above set, F_1, F_2, F_5 are associated with magnetic fields, F_3 F_6 with electric fields, and F_4 with both. (F_6 is the well-known Coulomb force used in electrostatic separators). F_1 and F_3 are second-order forces, in that they derive from dipole-type interaction with a non-uniform electromagnetic field; if the material is not ferromagnetic (in the case of a magnetic field) or reasonably conducting (in the case of an electric one), they are weak and call for higher intensities and gradients. F_2 and F_5 are due to interaction of the current and the magnetic field and can be used in eddy-current separation. F_4 is due to interaction of an electric and a time-variable magnetic field.

WHIES AND HGES: GENERAL BACKGROUND

The application of dielectric separation devices in the past has been limited by the practical difficulty of achieving a high value of electric field simultaneously with a high electric field gradient. This meant that an electric force large enough to permit filtration or separation by dielectric techniques was only developed on highly conducting materials.

However, it has recently been demonstrated (ref. 3,13) that simultaneous large values of electric field and electric gradient can be achieved over a large volume with an assembly consisting of electrodes surrounding a separation cell which contains a specially designed matrices.

High-gradient electric separation (HGES) and dielectric filtration (DF) are new techniques based on the use of the polarization force exerted by a non-uniform electric field (dielectrophoretic effects). Matrices of finely divided filamentary dielectric material and other types (glass fabric, titanates in the form of beads, balls or rods, perovskite-type ferroelectric bodies, ceramic wool, glass beads, pads, etc.) containing 50 to 95% void space are used. The system requires an electric cell capable of generating high-intensity fields in large volumes. The availability of these new types of matrices, whose elements are of the same dimensional order as those of the feed, extends the range of dielectric separation down to micron level.

Generally speaking, dielectrophoretic methods are suitable for: (1) cleaning conducting and non-conducting liquids of dielectric impurities; (ii) separating minerals in S-S-L or S-L systems, e.g. effecting concentration of finely divided cassiterite or removing precious metal grains from sands; (iii) upgrading precious metal concentrates; (iv) purifying oils and a variety of high polymers; and (v) particle analysis.

In the sequel, dielectric separation is discussed under the following aspects: (i) the intensity of the electric field and its gradient; (ii) size, density and dielectric constant of the particles; (iii) dielectric constant of the fluids serving as separation media; and (iv) the flow regime in the separator system.

There are several competing forces that act on a given particle in the separation system: dielectrophoretic, electric, gravitational, buoyancy and drag forces, the last-named resulting from the relative motion of fluid and particles. The London-van der Waals attraction force, double layer repulsion or attraction forces, and Brownian motion effects are normally negligible for suspended particles of diameter greater than about 5 microns.

The dielectrophoretic force acting on a small*, non-conducting, polarized but uncharged, spherical body in equilibrium, immersed in an insulating dielectric fluid and in a slightly non-uniform electric field, has been shown to be given by (ref.24).

$$F_e = \alpha V(E\cdot\nabla)E = \frac{1}{2}\alpha V\vec{\nabla}(E^2) \tag{9}$$

Where α is the polarizability, V the volume of the body, and E imposed external electric field. If the body is anisotropic, α is a tensor quantity. The computation then becomes more difficult.

For various shapes, α is given by:

$$\alpha = 3\varepsilon_1(\varepsilon_2-\varepsilon_1)/(\varepsilon_2+2\varepsilon_1) \; \text{....... for spheres}$$

$$\alpha = 2\varepsilon_1(\varepsilon_2-\varepsilon_1)/(\varepsilon_2+\varepsilon_1) \; \text{....... for long thin cylindrical rods}$$

$$\alpha = \varepsilon_1(\varepsilon_2-\varepsilon_1)/\varepsilon_2 \qquad \text{....... for a thin plate} \tag{10}$$

Equation (10) is valid only for a dilute suspension (ref. 25). For extension to higher concentrations see elsewhere (ref. 7). From equation (9) with $V = \frac{4}{3}\pi r^3$, the translational force, for perfect dielectrics**, is given by:

$$F_e = 2\pi r^3 \frac{\varepsilon_1(\varepsilon_2-\varepsilon_1)}{\varepsilon_2+2\varepsilon_1}\vec{\nabla}(E^2), \tag{11}$$

where r is the radius of the sphere, ε_1 and ε_2 are the absolute permittivities of, respectively, the surrounding non-conducting fluid medium and the particle, and $\vec{\nabla}(E^2)$ is the gradient that would be obtained at the specific location of the particles if they were not there. As F_e is a function of the gradient of the field strength squared, it is independent of electrode polarity and does not change direction under the influence of an a.c. field. F_e which is directed along the gradient of the electric field magnitude, depends also on the volume of the particle and on the mode of internal polarization. The effectiveness of dielectric separation decreases as the suspended particle size decreases.

---

* The dipole must be small compared to the scale of non-uniformities of the imposed electric field.

** Ones without conduction.

Since F_e increases with applied electric field strength, so does the separation efficiency. Equation (11), however contains no provision for conduction or frequency effects.

Any electric field, uniform or non-uniform, exerts a force upon a charged object. It is characteristic of non-uniform fields, however, that they exert a ponderomotive force upon small neutral objects (Figs. 1 and 2). Fig. 1 compares schematically the behaviours of neutral and charged bodies in a uniform and a non-uniform electric field, in air. Fig. 2 shows the case of dielectric levitation.

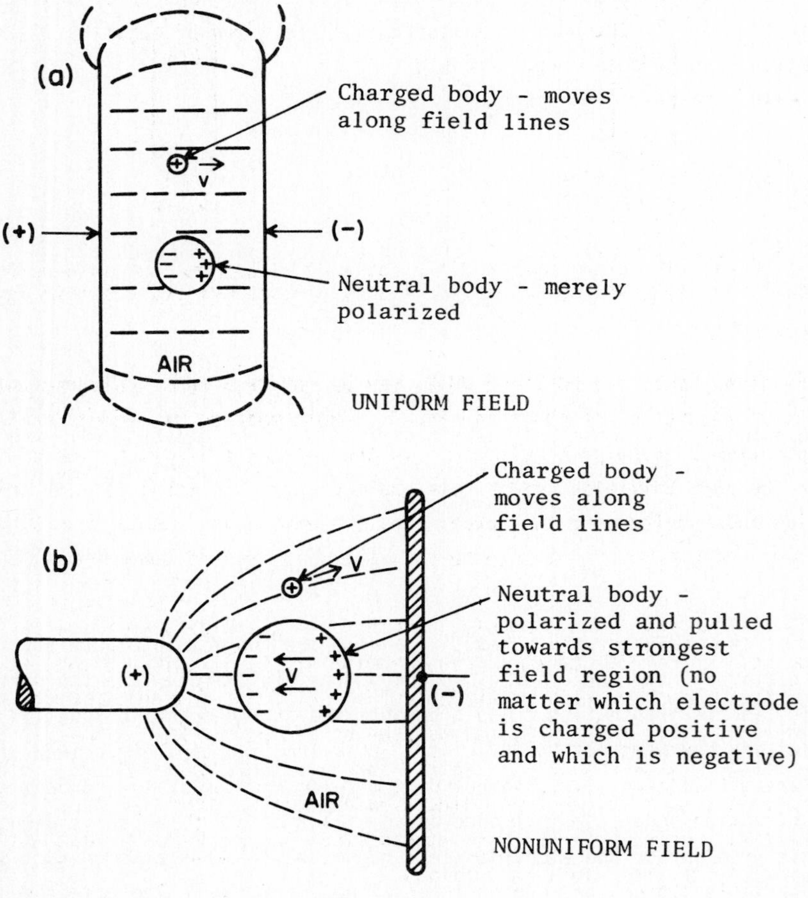

Fig. 1. Comparative behaviour of neutral and charged bodies in (a) a uniform electric field. (b) a non-uniform electric field (ref.14).

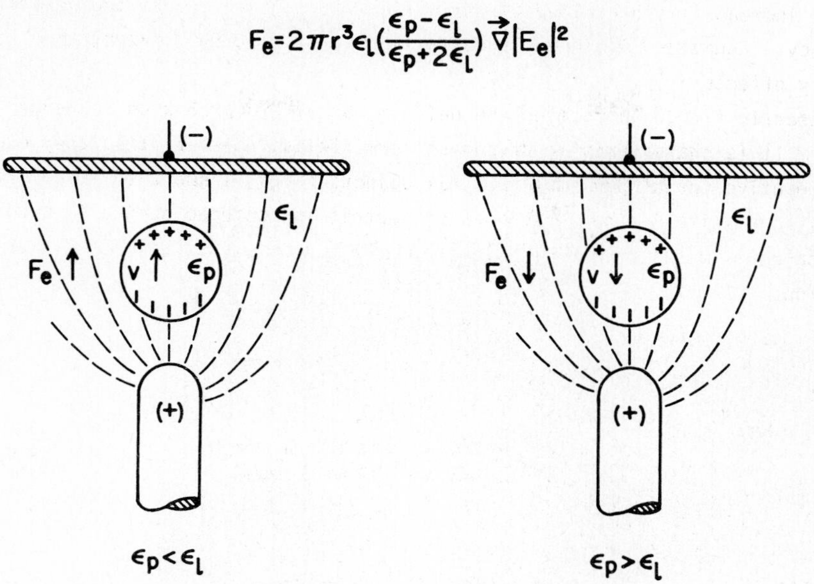

Fig. 2. Comparative behaviour of different neutral objects in a non-uniform electric field, immersed in insulating fluids, representing dielectric levitation.

On the other hand, the HGES and WHIES may be referred to as polygradient dielectric separators, as shown in Fig. 3. Both methods may serve for S-S or S-L separation of fine particulates. In the case of filtration, the difference between the permittivities of particle and medium, $\Delta\varepsilon$, determines both the direction and magnitude of the force, i.e. attraction vs. repulsion of the particles: when $\Delta\varepsilon = \varepsilon_p - \varepsilon_f > 0$, the particle is attracted towards the region of maximum electric field strength; when $\Delta\varepsilon < 0$, the particle is repelled into the region of lowest field strength. Note that F_e vanishes when the particle is suspended in a medium of equivalent permittivity, $\varepsilon_2 = \varepsilon_1$ in equation (11).

F_e is usually determined by the size and shape of the filament, and by the strength of the electrifying field, as seen by the matrix and particles.

If $\varepsilon_2 >> \varepsilon_1$ or $\varepsilon_2 << \varepsilon_L$, the dielectric force is independent of the dielectric constant of the particle, but its direction differs in the two cases.

If the particle is metallic, $\varepsilon_2 \rightarrow \infty$, and since the resistivity of the fluid is supposed to be very high, we obtain

$$F_e = 2\pi r^3 \varepsilon_1 \vec{\nabla}(E^2) \qquad (12)$$

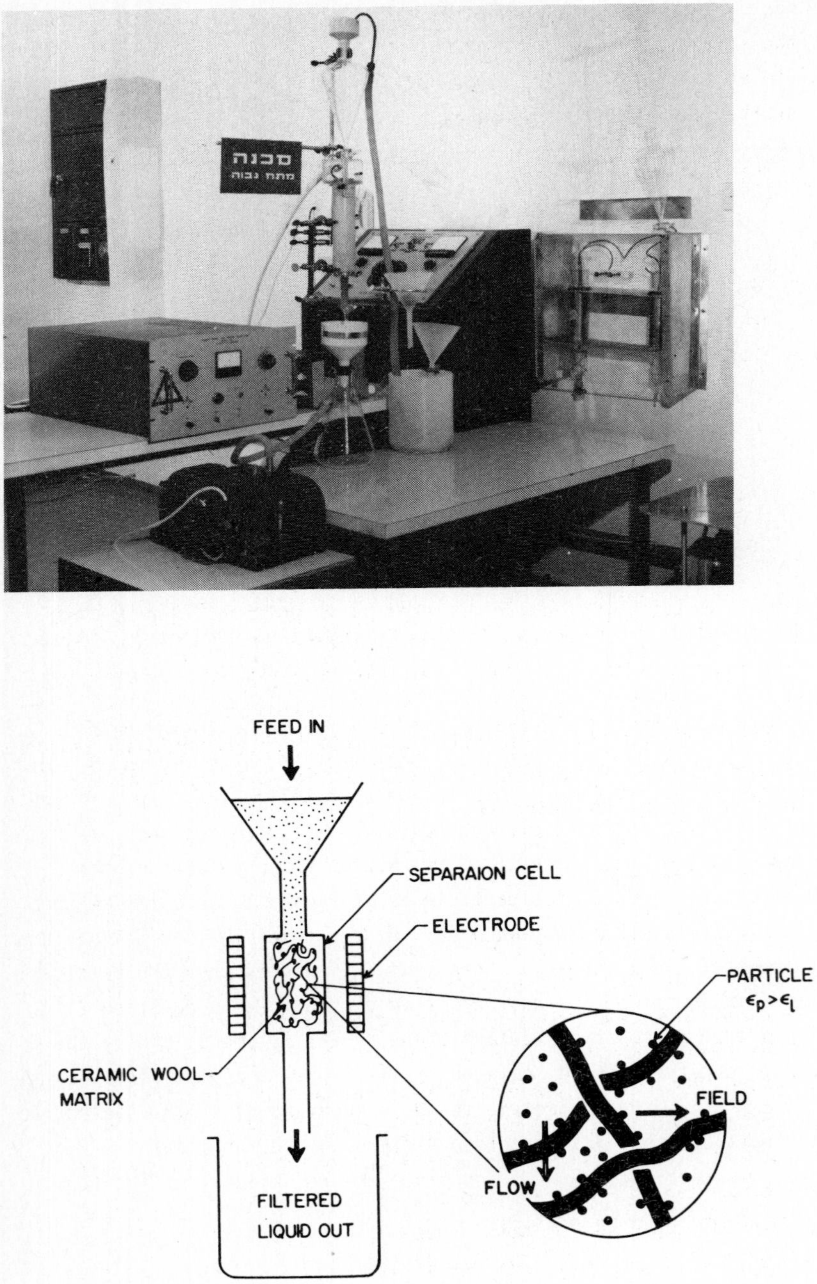

Fig. 3. Laboratory batch-type HGES unit.

Thus the force cannot be reversed and the particle will always move in the direction of the strongest fields.

For a particle of mass density, ρ_2, immersed in a dielectric fluid of density ρ_1, the net gravitational force is given by

$$F_G = \frac{4}{3}\pi r^3(\rho_2-\rho_1)\vec{g} \tag{13}$$

where $\vec{g}$ is the gravitational acceleration. This force is slight for small particles, but significant for larger ones.

The magnitude of the drag or viscous force encountered in this system will be, as predicted by Stokes' equation:

$$F_D = 6\pi\eta v_r r \ , \tag{14}$$

where η is the fluid dynamic viscosity, v_r the relative velocity between fluid and particle, and r the radius of the particle (assumed to be spherical).

The magnitude of the drag force determines the terminal velocity of the particles. The hydraulic force, F_D, which tends to strip a matrix filament of the particles adhering to it, is basically determined by the velocity of the fluid medium carrying the particles (the higher the velocity, the greater the stripping force).

The net force acting on a particle defines its trajectory, hence the equation of motion for a small spherical body (at vanishing particle Reynolds numbers) in a dielectric fluid can be approximated as

$$F_{net} = \frac{4}{3}\pi r^3 \rho_S \frac{dv}{dt} = F_e - 6\pi\eta v_r r \ , \tag{15}$$

where F_{net} is the inertial force and v is the particle velocity vector at time t. Equation (15) describes the trajectory of particles in the vicinity of the matrix and thus provides the necessary information for formulation of the criteria of capture (ref. 1).

Note that Brownian diffusion of small particles need not be taken into account, since it has no effect on the overall collection efficiency unless the dielectrophoretic forces are very weak. Other effects which may interfere, such as conduction and thermal convection, can also be neglected.

In conclusion, we may say that the main competing factors which determine whether or not a particle is trapped are the dielectric force, which attracts a dielectric particle to a matrix filament and holds it there, and the hydraulic force exerted by the fluid medium carrying the particle through the matrix (see Fig. 4). The filtration of $\varepsilon_p > \varepsilon_L$ particles from a fluid will occur when the ratio of the dielectrophoretic tractive force to the sum of the opposing forces is greater than 1.

Capture of particles by the matrix, particle trajectories, and the accumulation zone of the dielectric fraction, are discussed in detail by Lin et al (ref. 1). Note that Dielectrophoresis is to be distinguished from Electrophoresis, the motion caused by the action of an electric field (whether uniform or non uniform) upon a charged object.

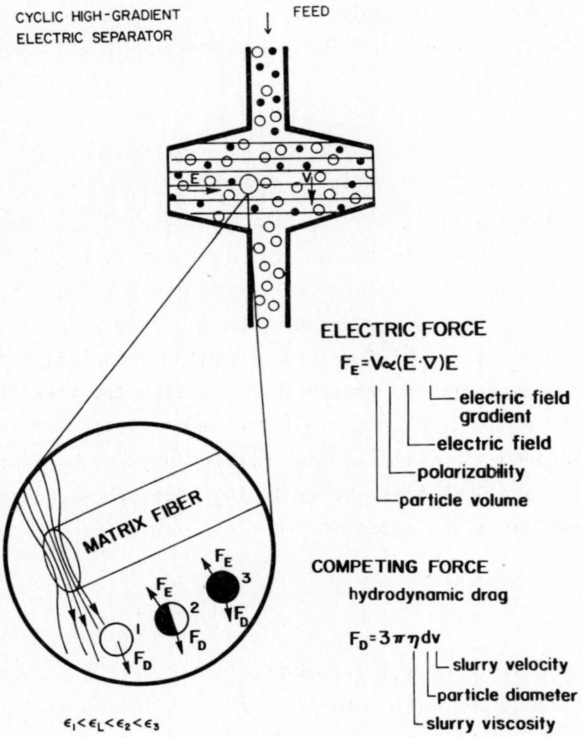

Fig. 4. Cyclic WHIES or HGES and the principal forces acting in the system.

DIELECTRIC CONSTANT OF A MIXTURE

The problem of determining the dielectric constant of a mixture of two materials with dielectric constants ε_1 and ε_2 is a very old problem (ref. 25). Various and numerous solutions have been given. It is a difficult problem and the fact that several solutions have been proposed indicates clearly that no one is completely satisfactory. The subject has recently received new interest (ref. 26) by using sophisticated mathematical methods, to try to determine rigorous bounds.

However, in the case of small concentrations satisfactory solutions have been found. We shall quote only a few of them which are particularly attractive because of their simplicity.

Consider particles of dielectric constant ε_2 embedded in a matrix of dielectric constant ε_1. The concentration of the medium 2 is C, and it is also supposed that the particles are spherical. Bottcher (ref. 27) gives the effective dielectric constant

$$\varepsilon = \varepsilon_1 + C \frac{3\varepsilon_1(\varepsilon_2 - \varepsilon_1)}{\varepsilon_2 + 2\varepsilon_1} \qquad (16)$$

For $\varepsilon_1 \simeq \varepsilon_2$ equation 16 reduces to $\varepsilon = \varepsilon_1(1-C) + \varepsilon_2 C$

A simple demonstration of this formula is given by Landau and Lifshitz (ref. 25). They consider the integral

$$I = \frac{1}{V}\int(D - \varepsilon_1 E)\, dV \equiv \overline{D} - \varepsilon_1\overline{E} \qquad (17)$$

where $\overline{D}$ and $\overline{E}$ are the mean values of the displacement D and the field E, in the composite medium. The integral is non-zero only inside the particles and it is proportional to the concentration C. Since the particles are supposed to be spherical, we can calculate the field E_2 inside a particle, assuming that the field around it is $\overline{E}$ (this is true only in the case of low concentrations). It is a standard result in electrostatics (ref. 25), that

$$E_2 = \frac{3\varepsilon_1}{\varepsilon_2 + 2\varepsilon_1}\overline{E} \qquad (18)$$

Thus we can calculate the integral I two ways:

$$I = \overline{D} - \varepsilon_1\overline{E} = (\varepsilon - \varepsilon_1)\overline{E} \quad (\text{since } \overline{D} = \varepsilon\overline{E}) \text{ and} \qquad (19)$$

$$I = \frac{1}{V}\int(D_2 - \varepsilon_1 E_2)\, dV = \frac{C(\varepsilon_2 - \varepsilon_1)3\varepsilon_1}{\varepsilon_2 + 2\varepsilon_1} \quad (\text{using } D_2 = \varepsilon_2 E_2) \qquad (20)$$

Equating the two results gives (equation 16). At low concentration we have a linear dependence of ε with C.

We can go a step forward using the Maxwell-Garnett theory (ref. 28). The result for the resultant permittivity, ε, is:

$$\varepsilon = \varepsilon_1 \frac{1 + 2C \dfrac{\varepsilon_2 - \varepsilon_1}{\varepsilon_2 + 2\varepsilon_1}}{1 - C \dfrac{\varepsilon_2 - \varepsilon_1}{\varepsilon_2 + 2\varepsilon_1}} \qquad (21)$$

It is easy to verify that for very small C (equation 21) reduces to (16). Finally, we mention the result of the Effective Medium Theory, which is valuable for all the range of concentrations $0 \leqslant C \leqslant 1$,

$$C \frac{\varepsilon_2 - \varepsilon}{\varepsilon_2 + 2\varepsilon} + (1-C) \frac{\varepsilon_1 - \varepsilon}{\varepsilon_1 + 2\varepsilon} = 0 \qquad (22)$$

Derivations of equations 21 and 22 can be found in reference 29.

For practical applications, equation 16 is very useful. As an example, we calculate ε for a mixture of two media with $\varepsilon_1 = 2$ and $\varepsilon_2 = 15$. We get that for $C = 0.05$, $\varepsilon = 2.21$ and for $C = 0.1$, $\varepsilon = 2.41$. These are relatively slow variations of ε with C.

DIELECTROPHORETIC FILTRATION OF NON-CONDUCTING LIQUIDS (REF. 11)

The Dielectric Forces

The force acting on a spherical particle of finite conductivity σ_2, and permittivity ε_2, embedded in fluid of finite conductivity σ_1 and permittivity ε_1, due to a non homogeneous and time dependent field, $E(x,y,z,t)$, has been calculated by Molinari and Viviani (ref. 30) to be

$$F = 4\pi R^3 \varepsilon_1 [E(t,x)*f(t)] \nabla E(t,x) \qquad (23)$$

In equation 23, * denotes the convolution product:

$$E(t)*f(t) = \int_{-\infty}^{t} E(t)f(t-\tau)d\tau$$

and the function f is given by

$$f(t) = \frac{\varepsilon_2 - \varepsilon_1}{\varepsilon_2 + 2\varepsilon_1} \delta(t) + 3e^{-t(\sigma_2 + 2\sigma_1)/(\varepsilon_2 + 2\varepsilon_1)} \frac{\varepsilon_1\sigma_2 - \varepsilon_2\sigma_1}{(\varepsilon_2 + 2\varepsilon_1)^2} U(t) \qquad (24)$$

with $\delta(t)$ the Dirac function and $U(t)$ the unit step function. If we suppose that the functional dependence of E with x and t is of the form $E = E_1(t)E_2(x)$, we get:

$$F=2\pi R^3 \varepsilon_1 [E_1(t)*f(t)]E_1(t)\nabla E_2^2$$

or

$$F=2\pi R^3 \varepsilon_1 g(t)\nabla E_2^2$$

with

$$g(t) = [E_1(t)*f(t)]E_1(t) \tag{25}$$

Thus the force F depends on

(a) The function g(t) which appears explicitly through the properties of the fluid and of the particle (through ε_2, ε_1, σ_2, and σ_1).

(b) The field gradient

Influence of the Material Properties

We calculate the function $g(t) = [E_1(t)*f(t)]E_1(t)$ in the case of a dc field, i.e. $E_1(t)$ is constant and equal to 1. We get

$$g(t) = \frac{\varepsilon_2 - \varepsilon_1}{\varepsilon_2 + 2\varepsilon_1} + 3\,\frac{\varepsilon_1\sigma_2 - \varepsilon_2\sigma_1}{(\varepsilon_2 + 2\varepsilon_1)\quad(\sigma_2 + 2\sigma_1)}(1 - e^{-t/\tau}) \tag{26}$$

with $\tau=(\varepsilon_2+2\varepsilon_1)/(\sigma_2+2\sigma_1)$. This quantity is the relaxation time characteristic of the system. If $t \ll \tau$, then $g(t) = (\varepsilon_2-\varepsilon_1)/(\varepsilon_2+2\varepsilon_1)$ and the conductivities σ_1 and σ_2 do not appear in the force, as in the case where $\sigma_1=\sigma_2 = 0$. Now if $t \gg \tau$, then $g(t)=(\sigma_2 -\sigma_1)/(\sigma_2+2\sigma_1)$ and in this case the force does not depend on the dielectric constants ε_1 and ε_2. We see that depending on the relative values of ε_1 and ε_2, σ_2 and σ_1, we can have very different values of the force for $t \ll \tau$ and $t \gg \tau$. In particular, it is possible that the force changes its direction with time as, for example, if $\varepsilon_2>\varepsilon_1$ but $\sigma_2<\sigma_1$. For more details see Ref. 11 and 12.

Some typical values for the electrical conductivity of common materials, at room temperature are given in Tables 4 to 6 and Figure 5. Table 5 gives a tabulation of some physical properties of several pure metals and alloys of interest. For typical alloys, in particular casting alloys, the conductivity is often substantially lower.

The Field Gradient

The determination of the field gradient in such a system is not directly possible. Without glass balls we can easily compute it, but the introduction of the balls seriously perturbs the field distribution and hence the field gradient. We are forced either to calculate it using a specifying model or to find an analogous system on which we can do measurements. We chose the second

TABLE 4

Electrical Conductivity of some materials at Room Temperature

Material	Conductivity $(\Omega.m)^{-1}$	Material	Conductivity $(\Omega.m)^{-1}$
Metals:		Ge (pure)	30
Cu	5.9×10^7	Sic (dense)	10
Mo	1.9×10^7	Fe_3O_4	10^4
W	1.8×10^7	boron carbide	200
Fe	1.0×10^7	Insulators:	
CrO_2	3.0×10^6	SiO_2, glass	$<10^{-12}$
ReO_3	5.0×10^7	Low-voltage porcelain	$10^{-10} - 10^{-13}$
Semiconductors:		Steatite porcelain	$<10^{-12}$

TABLE 5

Material Parameters of Several Metals and Alloys at Room Temperature

Metal	σ $(10^6 \Omega^{-1} m^{-1})$	ρ $10^3 (kg\ m^{-3})$	σ/ρ $10^2 m^2 \Omega^{-1} kg^{-1}$
Al	35.4	2.70	130
Mg	23.0	1.74	130
Cu	59.1	8.93	66
Ag	68.1	10.49	65
Zn	17.4	6.92	25
Yellow brass	15.6	8.47	18
Sn	8.8	7.29	12
Pb	5.0	11.34	4.5
Stainless steel (302)	1.4	7.9	1.8

TABLE 6

Relative permittivity and electrical conductivity of various liquids, at room temperature

Liquid	Relative permittivity ε	Electrical conductivity $\sigma(\Omega \cdot cm)^{-1}$
Hexane	1.80	6×10^{-18}
Heptane	1.92	2×10^{-17}
Carbon tetrachloride	2.24	5×10^{-16}
Benzene	2,28	5×10^{-16}
Toluene	2,38	6×10^{-15}
Isoamylether	2,82	2×10^{-14}
Diethyl carbonate	2,82	5×10^{-12}
Propionic acid	3.44	2×10^{-10}
Isoamylacetate	4.63	3×10^{-11}
Bromobenzene	5.40	2×10^{-12}
Propylacetate	5.69	4×10^{-9}
Tetrachloroethane	8.2	3×10^{-10}
Acetone	20.7	2×10^{-9}
Ethanol	24.3	7×10^{-9}
Ethylene glycol	38.9	3×10^{-7}
Distilled water	80.1	1×10^{-8}

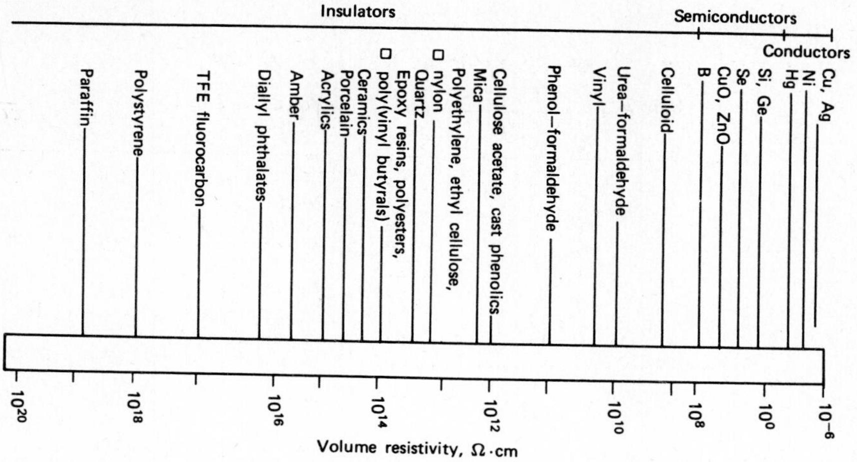

Fig. 5. Resistivity of materials

possibility in an electrolytic tank. The analogy is not complete for the following reasons: (a) The real system is three dimensional and the analogous one is two dimensional. (b) The real system is disordered, and it is difficult to realize the analogousness of disordered systems. Nevertheless, we think that the results obtained are very indicative and can take them as a qualitative determination of the field gradient.

In the electrolytic tank we measure the voltage distribution of a system made of two electrodes (two concentric circles). In the space between the electrodes we place stainless steel disks (which are analogous to the glass balls) in a close, compact arrangement and we fill the free space between the two electrodes with ordinary water (Fig. 6). This is an electrodynamic simulation of our real electrostatic system (all the relative dimensions are similar to the real system). Using a probe, we measure the voltage at several points in the interval between three contigous disks. As explained by the authors elsewhere (ref. 11), from these measurement we get an analytic expression V(x,y) for the voltage in this interval. Then we calculate the quantity $F_o = [(\partial E^2/\partial x)^2 + (\partial E^2/\partial y)^2]^{1/2}$ which is the absolute value of ∇E^2. We plot this quantity for several points located along some chords in the space between the electrodes (Figs. 7a, 7b, 7c). In the same figures we give the calculated variation of $F'_o = (dE^2/dr)$ for the case without discs. One sees that the experimental points are near the continuous calculated lines. But the important point is the ratio

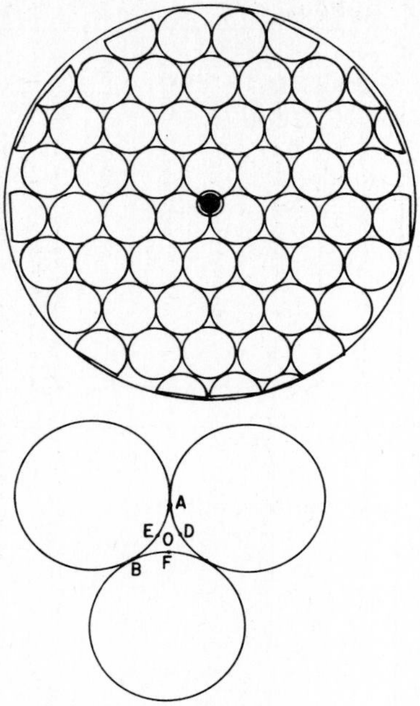

Fig. 6. Disposition of the stainless steel discs in the electrolytic tank.

of the measured values (with discs) to the calculated ones (without discs). For points located near the central electrode (Fig. 7a), this ratio is 100-150, but for the remote points (Fig. 7b) this ratio is near 500-1000. Thus we have a strong enhancement of the force, and it becomes larger and larger when one goes toward the outer electrode. This means that in this system the introduction of discs (or balls in the real system) gives a better field distribution than without discs from the point of view of getting large gradients in the whole apparatus. However, we have to remember that the values we get for the enhancement of the force is only for the analogous system. The conductivity ratio (between stainless steel and water) is above 10^3, and in our apparatus the permittivity ratio is only 3. The enhancement in the real system is certainly very much lower. We conclude that in the real system there is a small increase of the force, especially near the outer electrode.

In the above results the dielectrophoresis force was calculated at the point O (see Fig. 6) at the center of the triangle made by the three contact points of three discs (points ABC in Fig. 6). If we now compare the values of the force at these three points, we find that in the large majority of cases the force is larger near the points ABC than that at point O. This means that a particle

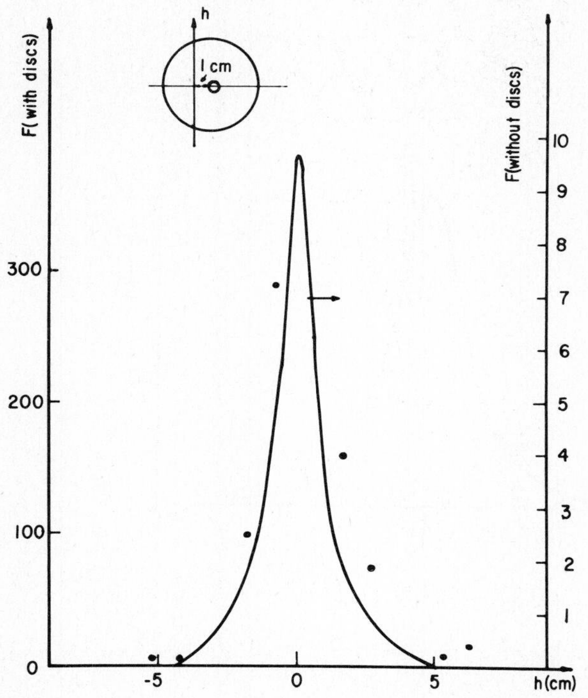

Fig. 7a. Absolute values of ∇E^2. The points are obtained from measurements in the electrolytic tank. The line is calculated and corresponds to the case without balls along a chord located 1 cm from the center.

will almost always drift toward the balls and be trapped if the force is such that its direction is toward the strongest field gradient (i.e. $\varepsilon_2>\varepsilon_1$, at $t<<\tau$ or $\sigma_2>\sigma_1$ at $t>>\tau$). But if the direction of the force is such that the particle is drifting toward the lowest value of the field gradient, it will move to point O and the particle will not be trapped.

Some Experimental Results

Fig. 8 shows a laboratory set-up for the fractionation of solids and the filtration of liquid by the HGES and DF techniques. The basic part is a cylinder (diameter 5 cm, length 27 cm) which constitutes the outer electrodes. The cylinder is filled with glass balls. We used two kind of balls with diameter 3 and 6 mm respectively. The inner electrode is connected to the high voltage power supply - while the outer electrode is grounded. In this work, we used

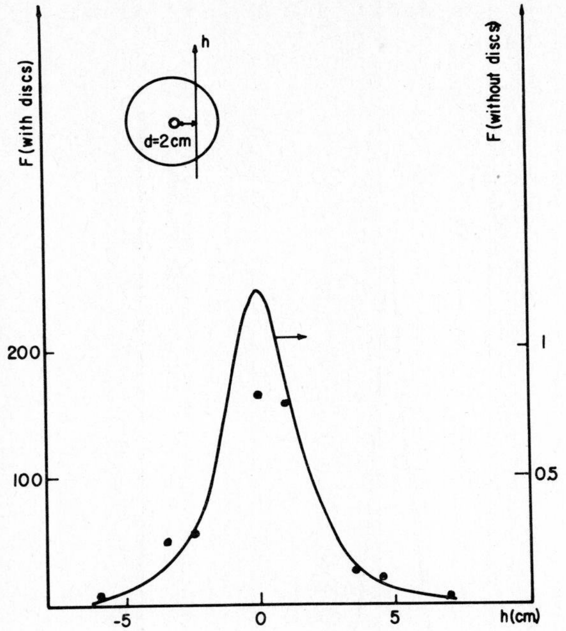

Fig. 7b. As for Fig. 7a but along a chord 2 cm from the center.

only a non-conducting liquid kerosene (heating oil). The liquid with a known quantity of powder flows through the main cylinder. In order to know the quantity of powder which is not trapped, the liquid is immediately filtered, as indicated in fig. 9. The quantity of trapped powder is recovered by rinsing the cylinder several times with oil, without applying voltage. Below we shall show that the necessary condition for the particle trapping is $\varepsilon_1 < \varepsilon_2$. In our case, $\varepsilon_1 \simeq 2$. The permittivities of the different powders are given in Table 7 with other properties. All the permittivities are larger than 4 and consequently we observe trapping for all these powders. The yield is clearly a function of several parameters: (a) The properties of the powder, such as permittivity, conductivity, density and grain size (see Table 7). (b) The flow rate: In the majority of cases we used a flow rate of 18 cm^3/sec. Since the volume of liquid was 350 cm^3, one experiment takes approximately 20 sec. However, to check the influence of the flow rate, we have also investigated a flow rate of 30 cm^3/sec. (c) The powder concentration: In this work, we were interested only in low concentration. As shown below, the yield is not depen-

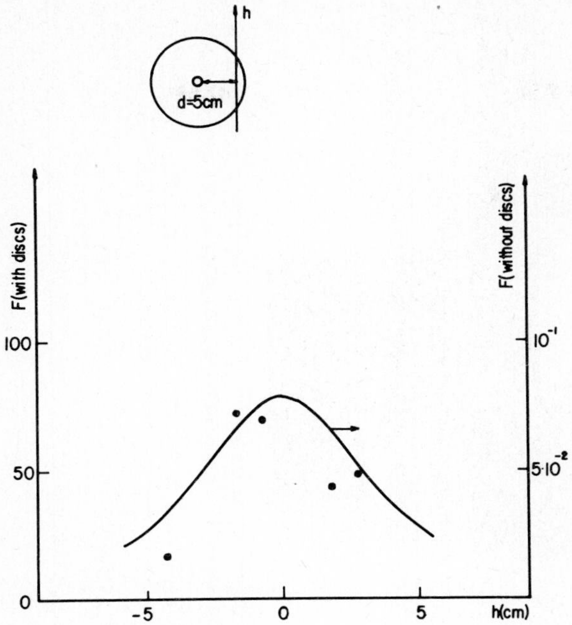

Fig. 7c. As for Fig. 7a but along a chord 5 cm from the center.

TABLE 7*

Material	ε	ρ(g/cm³)	R(μ)	v_L(cm/s)	g_0	τ(s)	$L=v_T\tau$	τ_c
MgO	9	3.6	35	0.68	0.54	3.5	8,5	11
Ilmenite	33	4.7	44	1.5	0.79	12	27	8
PVC	4.5	1.3	50	0.24	0.23	2	4	13.6
Cu		8.95	35	2	1		27	7.2

* ε(oil)=2; ρ(oil)=0.82 g/cm³

dent on the concentration in the range of interest. (d) The characteristics of the apparatus: Length, radius of the outer and inner electrodes. (e) The characteristics of the glass balls: Permittivity (5.5) and radius of the balls. As already mentioned, we found almost no difference using the two kinds of balls of different radii.

In Fig. 10, we present typical curves of the yield versus applied voltage, V, for MgO powder at different concentrations. At V=0, we have mechanical yield.

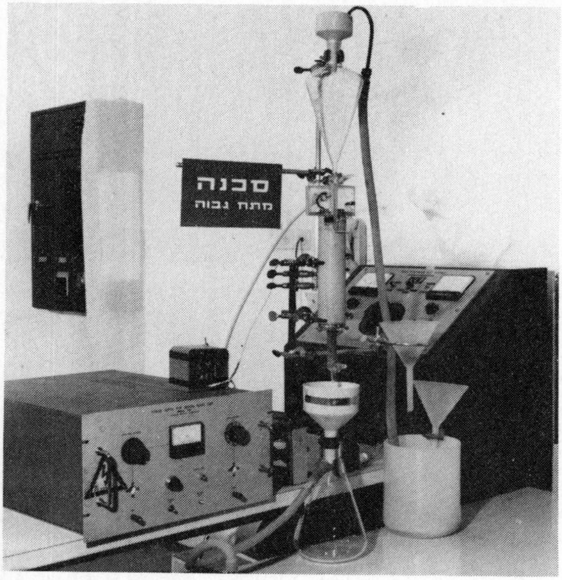

Fig. 8. General view of the filtration set-up.

It can be an important fraction of the total yield (from 15 to 40%). Although, we have not investigated this point, it can be an interesting advantage of the filtration system. By varying the voltage we first have a rapid increase of the yield and then a linear increase with V (with a slow slope). We note that the three curves of Fig. 10 are practically parallel, differing only by the values of the mechanical yield. This permits us to conclude that the electric yield is independant of the powder concentration (in the range of concentration chosen, below 0,5%).

In Fig. 11, we give the electric yield (total yield minus the mechanical yield) for PVC, MgO, Ilmenite and Copper powders. The shape of the curve is the same for all the cases. We shall discuss below the differences between these four curves as a function of the properties of the powders. At this stage, we note that with this apparatus we get good yields for very different materials. The fact that they are below or of the order of 60% shall not to be considered as a failure, since the filtration can be repeated several times. To check this point, we let the liquid with 2 gr of MgO flow through the apparatus twice: the total yield was 70% in the first flow and reached 98% at the second flow. We have investigated the yield as a function of the radius of

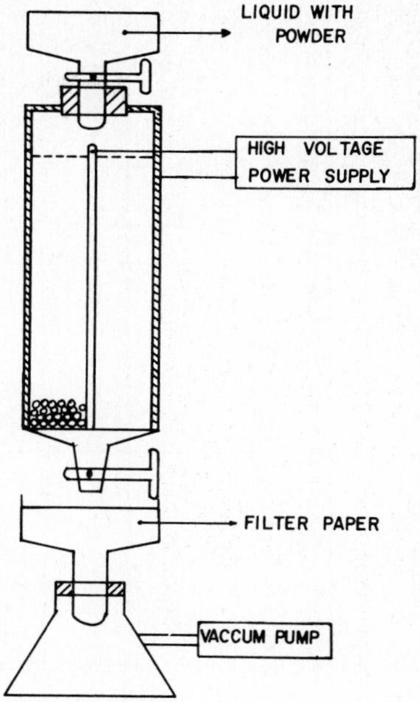

Fig. 9. Schematic view of the laboratory DF apparatus.

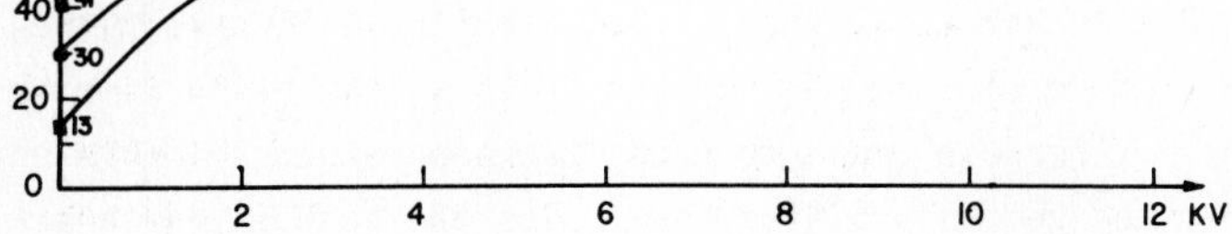

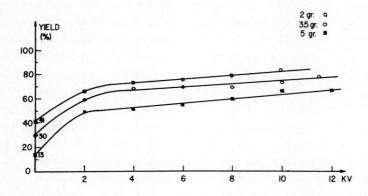

Fig. 10. Yield versus applied voltage for different MgO powder concentrations. Diameter of the balls: 3 mm. Diameter of the inner electrode: 8 mm.

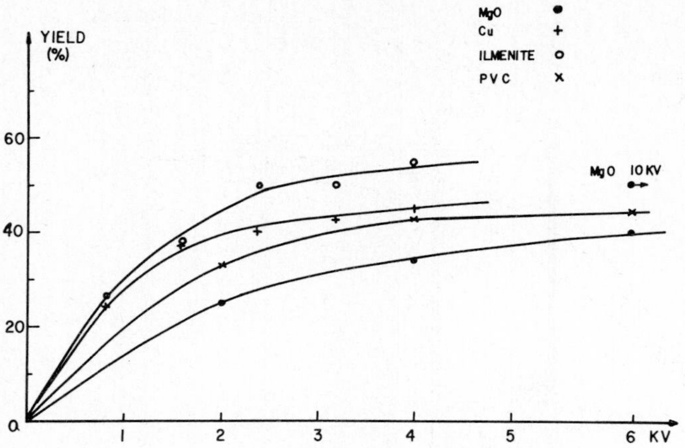

Fig. 11. Electrical yield versus applied voltage for different powders. Diameter of the balls: 3 mm. Diameter of the inner electrode: 8 mm.

the inner electrode and of the flow rate. We varied the radius from 2 to 6 mm and found (Fig. 12) that at large voltage (6kV) there are only minor variations of the yield with the radius. In Fig. 13 we have plotted the yield versus voltage for two different flow rates. As expected, increasing the flow rate from 18 to 30 cm^3/sec., decreases the yield significantly.

Interpretation of the Results

In this section we intend to discuss the results presented in Fig. 11 in order to understand the variations of the yield with the materials. We saw above that for $t<<\tau, g=(\varepsilon_2-\varepsilon_1)/\varepsilon_2+2\varepsilon_1)$ but for $t>>\tau$ we have $g=(\sigma_2-\sigma_1)/(\sigma_2+2\sigma_1)$. The only situation for which these two expressions are identical is in the case of a metallic particle (here, copper). In the other case we have a relaxation time equal to $(\varepsilon_2+2\varepsilon_1)/(\sigma_2+2\sigma_1)$. If we suppose $\sigma_1>>\sigma_2$ (for example, for PVC $\sigma_2\approx 10^{-12}\Omega^{-1}m^{-1}$ and for oil $\sigma_1\simeq 10^{-10}-10^{-11}\Omega^{-1}m^{-1}$) we get (depending on the material), $\tau\approx 2$-12 s. For $t\lesssim\tau$, since $\varepsilon_2>\varepsilon_1$, trapping will take place, but for $t\gtrsim\tau$, since $\sigma_2<\sigma_1$, there will be no trapping. This relaxation time has to be compared to the time for one grain to cross the apparatus. The velocity of a grain is equal to the velocity of the fluid plus the velocity of a grain relative to the fluid. We evaluate the velocity of the fluid as equal to 1.75 cm/s. For the relative velocity we suppose that the flow is not turbulent and we use the Stokes expression for the velocity

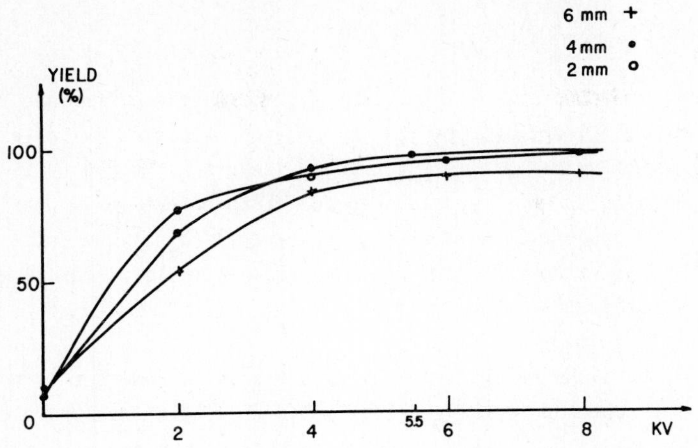

Fig. 12. Yield versus applied voltage for different inner electrodes with various diameters. MgO powder. Diameter of the balls: 6 mm. Note the larger values of the yield compared to Fig. 10. This is due essentially to the smaller radius of the inner electrode.

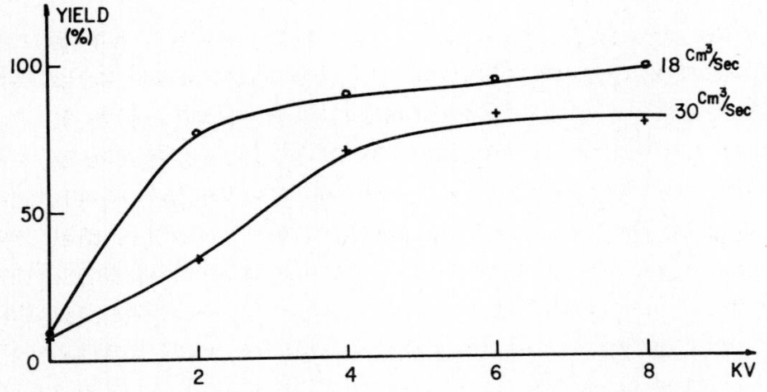

Fig. 13. Yield versus applied voltage for different flow rates. MgO powder. Diameter of the balls: 6 mm. Diameter of the inner electrode: 2 mm.

$$\nu_L = \frac{2}{9} \frac{R^2 g}{\eta} \Delta\rho \qquad (27)$$

g is the gravitation constant, η the viscosity, and $\Delta\rho$ the difference in the densities. Taking $\eta \simeq 1$ cps, we get the values of ν_L reported in Table 7. Thus we can calculate the crossing time t_c. We see that for MgO and PVC, we have $t_c > \tau$ but for ilmenite $t_c < \tau$. We can also define an effective length $L = \nu_T \tau$ in which the force is proportional to $(\varepsilon_2 - \varepsilon_1)/(\varepsilon_2 + 2\varepsilon_1)$ (ν_T is the total velocity of the grain). The values of L are reported also in Table 7 (for Cu we take L = apparatus length).

In order to compare the results of Fig. 11, we have to remember three main factors: (a) The factor $g_o = (\varepsilon_2 - \varepsilon_1)/(\varepsilon_2 + 2\varepsilon_1)$; (b) The effective length L, or the relaxation time τ; (c) The difference $\Delta\rho$ between the density of the grain and that of the fluid.

First, we compare the results for Cu and illmenite for which the effective length is equal to the apparatus length and the factor g is large. This explains why we get the largest value of yield for these two materials. However, although we have g (illmenite)<g(Cu), we see from the Fig. 11 that the yield of illmenite is larger. We think that this can be explained by the density difference which is much lower for ilmenite than for Cu. It is clearly a favourable situation for particle trapping. PVC and MgO are characterized by $\tau < t_c$. This means that only for a time approximately equal to τ, the factor g is equal to its value at t=0. For time larger to τ, g is given by $(\sigma_2 - \sigma_1)/(\sigma_2 + 2\sigma_1)$, and since $\sigma_2 < \sigma_1$, g is negative, i.e. no trapping. However, we get good values of the yield, especially for PVC, for which g and τ are relatively small. This is possible only if the trapping is very easy when $t < \tau$. Here too we think that the density difference may offer the explanation. This density difference in PVC is very low, $\Delta\rho \simeq 0.48$, and for MgO it is $\Delta\rho = 2.78$. As above, a low value for $\Delta\rho$ is very favourable for the particles to be trapped.

This study shows that one can very easily remove impurities from liquid by dielectrophoresis. We saw that the yield is satisfactory for all types of impurities we investigated (metallic, ceramic, minerals and plastic particles). However, we restricted ourselves to non-conductive liquid. Even in this situation we noted that it is necessary to take into account the low electrical conductivity of the liquid in order to understand the trapping phenomenon.

DIELECTROPHORETIC FILTRATION OF CONDUCTING LIQUIDS

The performance of DF with fluids of very low electric conductivity was analysed in detail in the previous chapters, and very good yields (ratios of trapped material to total amount of impurities in the liquid) were obtained with different kinds of impurities (minerals, ceramics, plastics and metals).

On the other hand, it was established that trapping becomes more and more unlikely as the relaxation time τ of the system decreases. Since τ decreases, inter alia, with increasing σ_1, we extend here to liquids with high conductivity and study the performance of the apparatus under these conditions, including a detailed theoretical analysis. In the last section, a new version of the apparatus adapted for conducting liquids, is presented.

Theoretical Considerations

The dielectrophoretic force exerted on a small spherical particle of radius R by a suddenly applied DC field E, is given by

$$F = 2\pi R^3 \varepsilon_1 \left[\frac{\varepsilon_2 - \varepsilon_1}{\varepsilon_2 + 2\varepsilon_1} + 3\,\frac{\varepsilon_1\sigma_2 - \varepsilon_2\sigma_1}{(\varepsilon_2 + 2\varepsilon_1)(\sigma_2 + 2\sigma_1)}\,(1 - e^{-t/\tau})\right] \nabla E^2 \qquad (28)$$

At t = 0, the force equals:

$$F(t{=}o) = 2\pi R^3 \varepsilon_1 \, \frac{\varepsilon_2 - \varepsilon_1}{\varepsilon_2 + 2\varepsilon_1} \, \nabla E^2 \, , \qquad (29)$$

while at $t \to \infty$ it equals:

$$F(t \to \infty) = 2\pi R^3 \varepsilon_1 \, \frac{\sigma_2 - \sigma_1}{\sigma_2 + 2\sigma_1} \, \nabla E^2 \, . \qquad (30)$$

Note that the positive force is defined in the direction of increase of the field.

At the inlet of the apparatus, a particle is subjected to the t = o limit of F, and the force varies along the path according to Equation 28. If the transit time is very large compared to τ, this is equivalent to $t \to \infty$, and the final force is given by Equation 30. Accordingly, we write:

$$F(t{=}o) = 2\pi R^3 \varepsilon_1 g_o \nabla E^2 \qquad (31)$$

and

$$F(t{>>}\tau) = 2\pi R^3 \varepsilon_1 g_\infty \nabla E^2 , \qquad (32)$$

where $g_o = (\varepsilon_2 - \varepsilon_1)/(\varepsilon_2 + 2\varepsilon_1)$ and $g_\infty = (\sigma_2 - \sigma_1)/(\sigma_2 + 2\sigma_1)$,

Regarding the role of the glass-ball matrix (dielectric constant ε_b) in disturbing the field, it was shown (ref. 3,11) that given $\varepsilon_1 < \varepsilon_b$, the gradient is larger at contact points of the balls and smaller in the gaps between them.

If $\varepsilon_2>\varepsilon_1$ and $\sigma_2>\sigma_1$, both g_0 and g_∞ are positive and so is the force, irrespective of the time variable; thus the particles are attracted to the matrix and trapped. In the reverse case (both g_0 and $g_\infty<o$), the particles tend to drift away into the gaps in the matrix and trapping is prevented.

In fact, there is always, even in the absence of field, some trapping which we call the mechanical trapping. The trapping discussed above was in addition to the mechanical trapping which is set as zero of our measurements.

In this chapter we are concerned with conducting liquids, $\sigma_1>\sigma_2$, i.e. $g_\infty<0$. For filtration to be possible in these circumstances, it is thus necessary that $g_0>0$ or $\varepsilon_2>\varepsilon_1$ - a condition complied with throughout our experiments.

With $g_0>0$ and $g_\infty<0$, the force is positive for $t = 0$ and later goes through a change of sign at a time t_0. The latter is obtained by setting $F(t_0) = 0$ in Equation 28, namely

$$t_0 = \tau \ln \frac{3(\varepsilon_1\sigma_2-\varepsilon_2\sigma_1)}{(\sigma_2-\sigma_1)(\varepsilon_2+2\varepsilon_1)} \tag{33}$$

If $\sigma_2<\sigma_1$, but the two values are close (for example, $\sigma_2/\sigma_1 \simeq 0.9$), t_0 may be much larger than τ. If, however, $\sigma_2<<\sigma_1$ such that $\varepsilon_1\sigma_2<<\varepsilon_2\sigma_1$, t_0 equals:

$$t_0 \simeq \tau \ln \frac{3\varepsilon_2}{\varepsilon_2+2\varepsilon_1} \tag{34}$$

In general, we shall see that t_0 is slightly smaller than τ, but of the same order of magnitude. Note also that if $\varepsilon_2>>2\varepsilon_1$, $t_0 \simeq \tau$, and if $\varepsilon_2 = 2\varepsilon_1$, $t_0 \simeq 0.4\tau$.

We thus see that, denoting the total velocity of the particle by v, the dielectrophoretic force is positive over a path segment $L_1 = vt_0$, along which the particle is attracted by the matrix. The probability of actual trapping in this segment is substantial, although very difficult to determine. It depends on many parameters. The most important are the density differential between the impurities and the liquid, and the friction between them and the matrix. It should also be borne in mind that the dielectrophoretic force may be very small away from the inner electrode. By contrast, beyond t_0, along the rest of the path $L_2 = L - L_1$ (L being the total length of the apparatus), the probability of trapping is low, as the particle is repelled by the matrix. In other words, the closer L_1 to L, the more favourable the conditions for filtration. The situation was demonstrated by us in (ref. 11) on a mixture of kerosene and ilmenite, with very high yields under relatively low voltages (~4 kV).

It is now also clear why our apparatus becomes less efficient (i.e. the yield decreases as liquid conductivity increases): the relaxation time τ and inversion time t_0 decrease and consequently L_1 decreases while L_2 increases; eventually, with $L_1 \ll L_2$, the yield may drop below the mechanical trapping level.

All the above refers to insulating impurities. For metallic impurities $\varepsilon_2, \sigma_2 \to \infty$, and equations 31 and 32 reduce to the single expression:

$$F_{met} = 2\pi R^3 \varepsilon_1 \nabla E^2 \ . \qquad (35)$$

It is seen that the force is always oriented throughout in the direction of increase of the field. Conditions for filtration are always favourable ($g_0 = g_\infty > 0$) and the yield is not affected by the conductivity of the liquid. Equation 35 can be obtained directly from the well known results in electrostatics that a conducting sphere in a field is equivalent to a dipole equal to $4\pi R^3 \varepsilon_1$ and from the general expression of the dielectric force $F = p\nabla E$, one gets equation 35.

Experimental and Results

As in previous works (ref. 3, 7, 11, 13) we use the same liquid as before - kerosene (heating oil) with its conductivity (originally 10^{-10}-$10^{-11} \Omega^{-1} m^{-1}$) progressively increased through admixture of polar liquids such as dichloromethane (CH_2Cl_2) and isopropanol, in such percentages as to stay within the requirements $\varepsilon_2 > \varepsilon_1$ and $\varepsilon_1 < \varepsilon_b$. The variation of the dielectric constant (moderate) and conductivity (steep) with the percentage of polar admixture is shown in Figs. 14 and 15 respectively.

We have worked with two types of impurities: insulating and metallic. The insulating impurity used was PVC (dielectric constant of the order of 4.5), percentage 1%, and the metallic impurity - copper (0.18%). These low concentrations were adopted in order to avoid short-circuits by bridging, and to present a case for filtration of dilute suspensions.

Results (yields as function of applied voltage, Figs. 16 and 17) are in agreement with the theoretical analysis in the preceding section:
(a) For the metallic impurity no change of the yield was observed over a range of variation of σ_1 by three orders of magnitude. (The scatter of the experimental points is due to measurement errors). (b) By contrast, for the PVC the yield is seen to vary significantly with σ_1. As the dielectric constant of the kerosene-isopropanol mixture varies only moderately in the chosen concentration range, the variation of the yield cannot be attributed to the change in g_0. As explained above, the length L_1 decreases as σ_1 increases, and the apparatus is less efficient. (c) It is remarkable that for the 15% mixture we have a decrease of the yield as the voltage increases, below the level of its mechanical

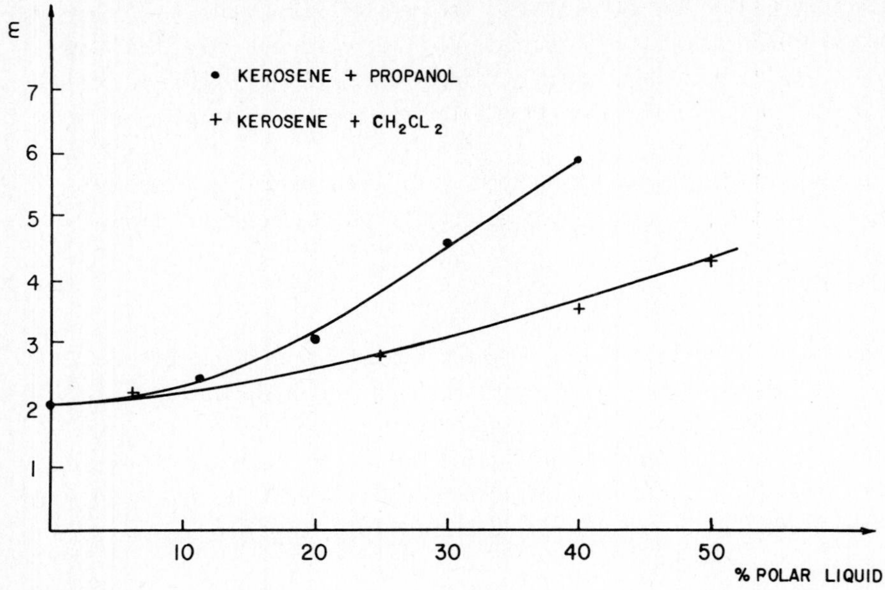

Fig. 14. Dielectric constant of the kerosene-polar liquid mixtures versus polar-liquid concentration.

counterpart.

The reason is that in this case L_1 is so small (~1 mm) that $L_2 \simeq L$. During the crossing of the apparatus, the particles are repelled by the beads and consequently the total yield is smaller than the mechanical yield.

In terms of relaxation time, the picture is as follows. Assuming $\sigma_2 << \sigma_1$ we get for pure kerosene $\tau \simeq 2$ sec and $t_0 \simeq 1$ sec, and since $v \simeq 2$ cm/sec we have $L_1 = 2$ cm and $L_2 = 25$ cm; yet the yields observed were very good (~98%). The conclusion is that apparently σ_1 and σ_2 are of the same order of magnitude. For the 10% mixture $\tau = 0.17$ sec, and taking $\sigma_2 \simeq 5 \times 10^{-11}$, $\sigma_1 = 2 \times 10^{-10} \Omega^{-1} m^{-1}$, we have $t_0 \simeq 0.5\tau$. L_1 is already small and the dielectric yield at 7kV is about 30% - by no means negligible. For the 15% mixture, however, τ decreases to 10^{-2} sec - far too short for dielectric trapping.

New Version of DF Apparatus

In the light of the above, there is no advantage, when $\sigma_2 << \sigma_1$ - in increasing the length of the apparatus. The answer lies, rather, in creating a repetitive pattern of the force as per equation 31 at $t = 0$. This was realized by us in

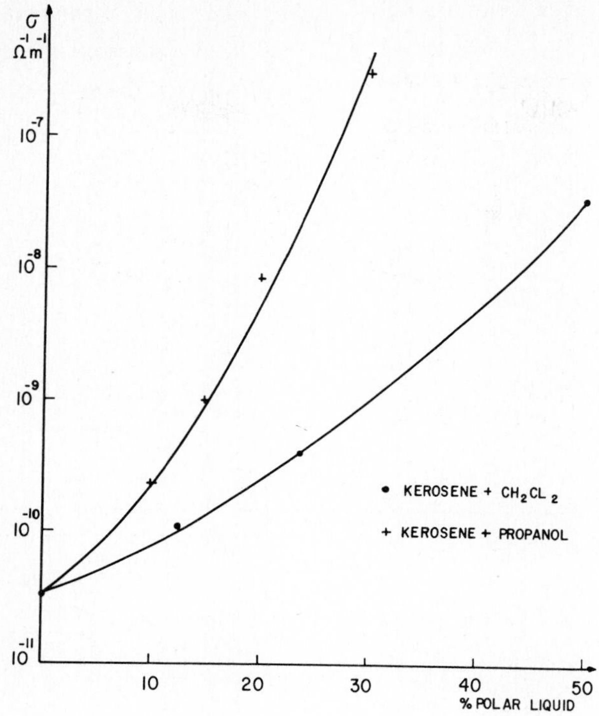

Fig. 15. Electric conductivity of kerosene-polar liquid mixtures versus polar-liquid concentration.

a three-stage apparatus (Fig. 18) where three active field zones 1,2,3 alternate with two zero-field zones A, B, all of which are crossed successively by the particle. As the stages are shorter (one-third, in this case) than the original apparatus, the operation is not equivalent to multiple repetition of the process in the original device. On the other hand, since the overall length and the flow rate are the same, the transit time also remains the same - but the trapping probability is tripled by recurrence of the $t = o$ conditions, and the repulsion probability is reduced. Thus an increase in the total yield can be expected with the same material. (Generally, the higher the conductivity of the liquid, the greater should be the number of stages and the shorter each stage).

The results are given in Fig. 19. For pure kerosene, the yield voltage is almost identical with that of Fig. 17, confirming that $\sigma_2 \simeq \sigma_1$. The improvement in performance is especially marked for the 15% mixture: the dielectric field is now positive, and the yield is 50% for 8 kV.

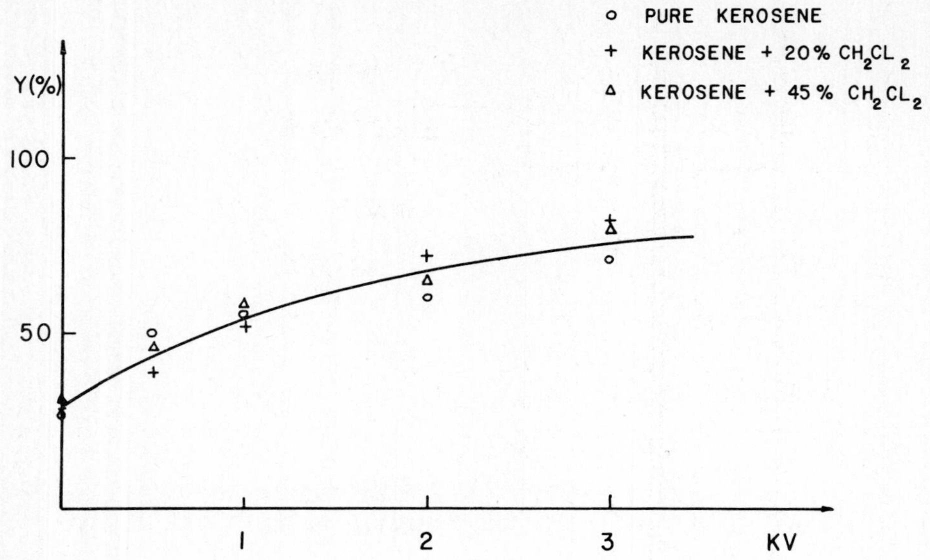

Fig. 16. Yield versus applied DC potential for copper particles.

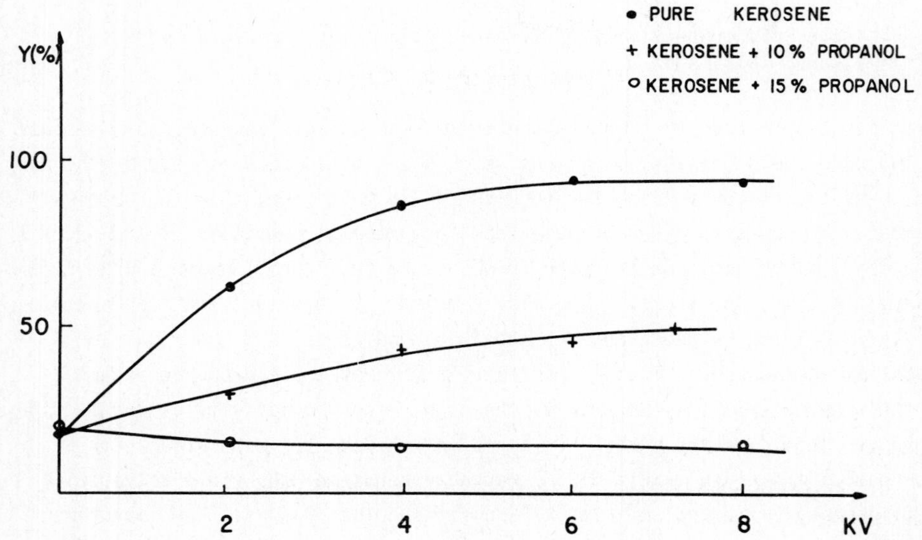

Fig. 17. Yield versus applied DC potential for PVC particles in conventional apparatus.

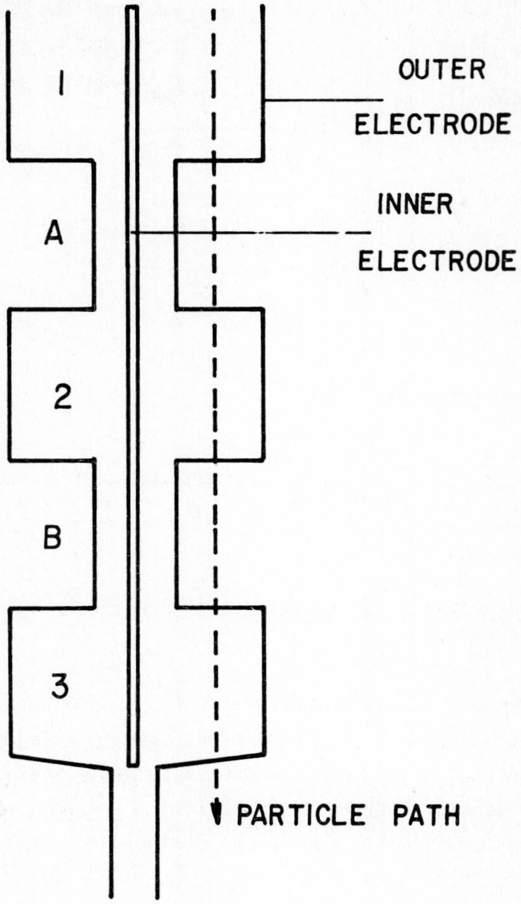

Fig. 18. Multistage filtration apparatus for conducting liquids. (Active field regions 1, 2, 3 filled with glass balls; zero field regions A and B empty).

A similar improvement was obtained for MgO ($\varepsilon \simeq 9$) in a mixture of kerosene and dichloromethane ($\sigma_1 \simeq 10^{-8}\Omega^{-1}m^{-1}$): at 6kV the yield was 32% in the original apparatus and 80% in the three-stage. The fact that the yield was larger for MgO than for PVC (compare Fig.15) is essentially due to the higher value of g_o in the former.

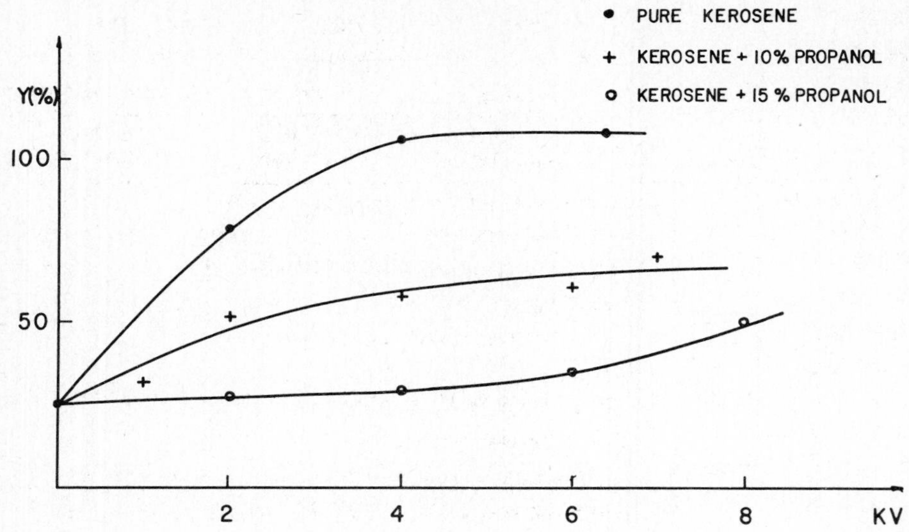

Fig. 19. Yield versus applied DC potential for PVC particles in three-stage apparatus (Compare Fig. 17).

To sum up, the experiments in dielectrophoretic filtration were in complete agreement with the theoretical analysis presented above. While for insulating impurities the yield decreases as liquid conductivity increases, in the case of metallic impurities it is independent of the latter. In the light of the results, a more efficient multistage device was designed and satisfactory yields were obtained for $\sigma_1 \simeq 10^{-7}\Omega^{-1}m^{-1}$.

MATRICES

The simplest way to create an electric field gradient is to introduce a dielectric body (such as ceramic beads) into the space occupied by a uniform electric field. The lines of force tend to concentrate in the body and generate a non-uniform field in its vicinity. Non-uniformity prevails only in the neighborhood of the matrix elements and vanishes in practice at distances comparable to the diameter. The field gradient is produced exclusively by the element, and its maximum value occurs at the surface of the element. The smaller the radius, the higher the maximum value of gradient. Therefore the way to realize high gradients is to employ a filamentary matrix (with elements of small radii of curvature).

High gradients are therefore uniquely suitable for handling fine particulates.

Maximization of the electric field gradients is feasible in HGES compared with WHIES units.

High-gradient matrix (random-oriented filamentary type) structures, sufficiently fine to handle colloidal particles, are also sufficiently open to permit very high flow rates. The matrices are able to retain several times their own weight in collected solids and release them instantly during the washing stage. With matrices consisting of smooth spherical ceramic perovskite compounds, glass beads or rods, etc., separation of electrically conducting materials may be very effective. This type of matrix causes the particles to collect preferentially at the points of contact, thus preventing them from bridging across the electrodes and shorting out the high voltage field.

A HGES or WHIES units which is clogged at least partially because of the retention of particulates in the interstices is usually referred to as "loaded unit". The response of a loaded device to further input depends on the nature of the geometrical changes in the flow passages introduced by prior capture and deposition. The following are the changes in the stucture and properties of the matrix:

- geometry of the matrix elements
- surface characteristics of the matrix elements
- the electric field distribution near the matrix surface
- the interstitial velocity of the fluid suspension
- the cross sectional areas available for flow
- the macroscopic porosity of the filter bed.

These changes, which are function of the time and the operating variables, reduce the DF, HGES, and WHIES performance (filtration rate, pressure drop, clarification or separation efficiency). The principles and difficulties of a theory of fibrous filter with electrical forces are essentially the same as those with magnetic forces.

For a given matrix configuration and electric field strength, the performance of an HGES device is determined by the velocity of the feed slurry through the matrix. The process is thus velocity-limited. The residence time of the slurry in the field is purely a dependent variable determined by the slurry velocity and the axial length of the matrix, the latter being such as to assure uniform flow distribution. The total process capacity of the HGES unit is roughly proportional to the flow cross-sectional area of the matrix. In both, HGES and DF increasing the electric field strength would allow a proportional increase in flow velocity for the same performance and consequently a proportional increase in capacity.

A filter of this type is very effectively cleaned by first de-energizing it and then back-flushing with sufficient vigor to cause the beads, spheres, or rods to agitate violently. This not only releases the trapped material but prevents any tendency to harden, adhere to the beads, etc.

If the permeability of the matrix is too low, the slurry should be fed at high pressure for economically acceptable flow rates to be achieved.

The "duty cycle" of a static or a cyclic HGES device is that fraction of the total operating time in which feed material passes through the matrix (with the electric field activated), expressed as percentage:

$$\text{Duty cycle} = \frac{\text{Feed Time}}{\text{Feed + Rinse + Flush Time}} \times 100 \qquad (36)$$

The performances of two units with different matrix length but otherwise identical will be approximately equal if the feed, rinse and flush volumes are all scaled directly proportional to matrix volume or, as in the present case, to matrix length. This scaling is also necessary if the feed, rinse and flush velocities are to be kept constant.

In addition to the dielectrophoretic trapping of particles there will of course be a significant fraction of particles that are mechanically trapped by the matrix elements. The mechanically trapped material will include both $\varepsilon_p<\varepsilon_1$ and $\varepsilon_p>\varepsilon_1$ particles. It is assumed that the mechanical capture efficiency is the same for both type of particles and that the mechanical capture increases with the filter length, filling factor of the matrix in the filter, and size of particles.

An alternating voltage applied to the electrodes produces the same basic results as a direct voltage, except that the former helps in preventing bridging and trapping of particles in the matrix by means of the jigging and shaking effects induced by it in the system, leading to better selectivity both in S-S and S-L separation.

Dielectric separation depends crucially on correct optimization of such parameters as matrix configuration, background field intensity, flow velocity, matrix loading, particle size and surface, chemical environment, colloidal stability of the particles, etc. Generally speaking, all operational variables are determined experimentally.

A strong electric field, although an essential part of the separation process, is in itself not sufficient; it must also exhibit a steep gradient. All three techniques, WHIES, HGES, and DF, customarily employ a matrix of shaped dielectric pieces in the separation zone, producing the necessary gradient and acting as collection sites. A variety of different types of matrices have been proposed (ref. 3, 13), including beads, balls, rods, fibers, grooved plates, grids, needles, pads, etc. The matrix should be chosen so as to best fit the characteristics of the material to be separated. In choosing a matrix, we have to consider the following:

- maximum electric field gradient
- volume-average electric field gradient
- collection surface area per unit volume of collection zone
- porosity and permeability of matrix to flow
- permittivity of matrix
- matrix size (electric force increases so long as matrix size exceedsparticle size).
- corrosion resistance.

Two main parameters which characterize any matrix of a given geometry are its size and porosity, and the ratio of void volume to total volume. The material used to form the matrix should be of high ε value such as perovskite ferroelectric barium titanate. $BaTiO_3$ proved to be ideal in applications because its electrical properties can be controlled within a wide range by means of mixed crystal formation and doping. For example (i) by substituting Ba^{2+} ions with Pb^{2+} ions, T_c can be increased linearly up to 490°C for a 100% Pb substitution (ii) in the same manner T_c can be continuously decreased by the substitution of Ba^{2+} with Sr^{2+} or of Ti^{4+} with Zr^{4+} or Sn^{4+} (see Fig. 20). (iii) The dielectric constant of $BaTiO_3$-ceramics has a strong temperature dependence and a pronounced maximum at the Curie point, as shown in Fig. 21. By a partial substitution of Ba by Sr or Ca or of Ti by Zr or Sn (Fig. 20), the Curie point can be shifted to room temperature, resulting in materials with maximum ε of 10^4 to $1{,}5 \times 10^4$ at this temperature (Fig. 21).

Owing to their ferroelectric nature, $BaTiO_3$-based ceramics show high dielectric constants and therefore these materials are suited for HGES and DF applications.

In the development of theoretical separation models, several common approaches and assumptions are generally used to obtain mathematical equations. The solid particle is assumed to be spherical, uniformly distributed in the fluid stream, and moving at the same velocity as the fluid passes the separator. The matrix is assumed to be of a given configuration and geometry. In reality, however, most matrices are characterized by random packing arrangements, which makes rigorous mathematical analysis of the problem very difficult. In these circumstances, we have to confine ourselves to regular arrays.

A major objective is designing a DF or WHIES unit is to ensure that it will provide the desired performance at minimum cost. Generally, this means that the ratio of mass transfer rate to the pressure loss should be maximized for the system. In the case of packed beds, the geometry or structural parameter used in correlating transport phenomena is the mean packing voidage. It is known that different ordered packings of spheres having identical mean voidage can have very different pressure losses. (Ref. 32). Since transfer is related to system geometry we have to predict geometries that yield optimum transfer to

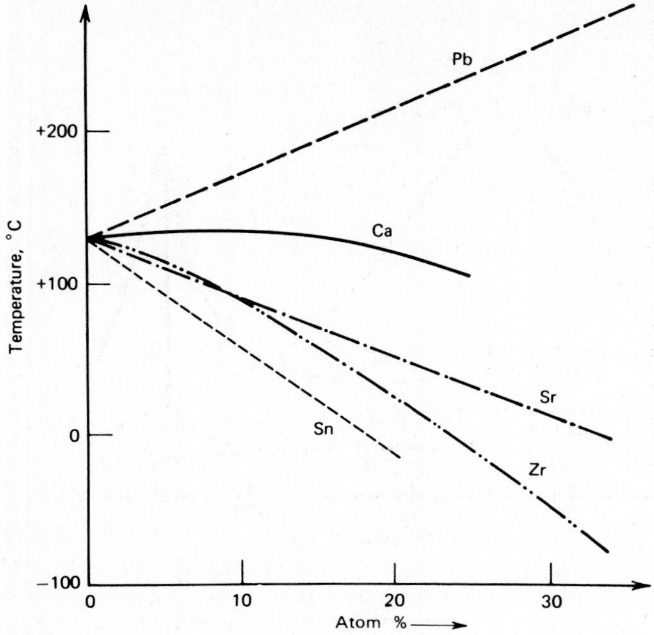

Fig. 20. Shift of the Curie point of $BaTiO_3$ by substitution of Ba^{2+} ions by Pb^{2+}, Ca^{2+}, or Sr^{2+} ions, or of Ti^{4+} ions by Zr^{4+} or Sn^{4+} ions (Ref. 31).

drag ratios.

Fig. 22 shows, in two views, some possible packing configurations of balls that may be stable to a certain extent under the action of filling of a given canister. Table 8 lists some of the properties of such arrangements. It is seen that the initial porosity (at operation-time zero) ranges from 25.95 to 47.64% and the number of ball contacts* - from 6 to 14.

In a laboratory WHIES device with a cylindrical canister, filled with random-packed glass balls, the porosity ranges from 34 to 40% (dense and loose arrangements respectively), with a mean of 7.1 ± 0.5 contacts per ball. Estimation of the average number of contacts is important for evaluating the separation efficiency, since F_e is maximized at these points and near them. For the case of non-uniform size, see elsewhere (Ref. 33).

*Obtainable by filling the canister with hot tap water, and then draining the water and drying the system.

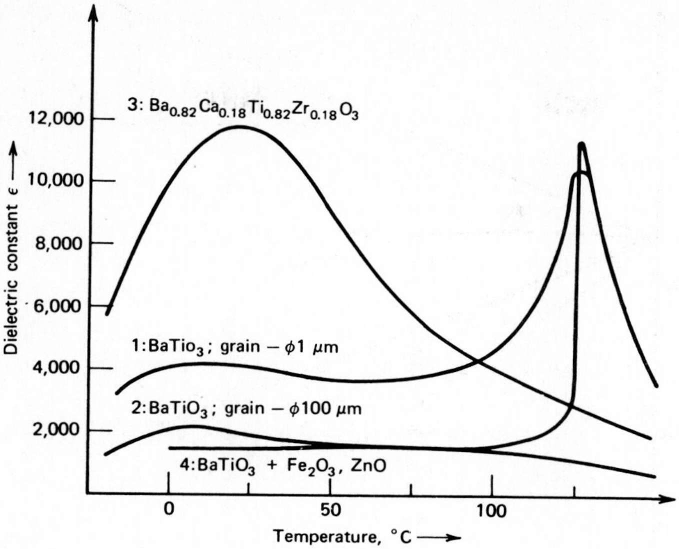

Fig. 21. Dielectric constant vs temperature of different $BaTiO_3$-based ceramics for capacitor applications (Ø = mean grain diameter).

TABLE 8

Properties of Different Types of Regular Packing Arrangement of Balls

Packing	Type of Polyhedron	Initial Porosity %	Number of Ball Contacts
Case 1 Cubic	Cube	47.64	6
Case 2 Orthorhombic (I)	Hexagonal prism	39.54	8
Case 3 Rhombohedral (I)	Dodecahedron	25.95	12
Case 4 Orthorhombic (II)	Hexagonal prism	39.54	8
Case 5 Tetragonal-Sphenoidal	Fourteen-sided polyhedron	30.18	14
Case 6 Rhombohedral (II)	Dodecahedron	25.95	12

Factors tending to reduce the porosity (and thereby the separation efficiency) are: (i) accumulation of $\varepsilon_p > \varepsilon_f$ particles on the matrix surface, which increases with time of operation, (ii) mechanical trapping of particles* , and (iii) the fluid flow rate and pressure drop across the matrix. A number of models have been presented for the capture mechanism of particles by the matrix elements. The first of these (Ref.1) is based on the work of Zebel (Ref. 34) for an electrostatic collection system. HGES and DF involve impingement and attachment of particles of varying size and dielectric constant on fine dielectric fibers placed in a uniform background field. In a practical HGES filter, a filamentary matrix has a packing fraction of about 5%. (Such a packing fraction, with large surface-to-volume ratio, is essential for the case of gaseous carrier). Our own basic model (Ref.1) refers to capture of small dielectric particles on filamentary elements and involves calculation of the particle trajectories near the latter. In all models, we have to consider the fraction of fibers active in the capture process.

Several models of the HGES and DF processes have been developed, but the design and performance characterization of practical systems are still largely empirical.

From our tests, it is clear that particle collection occurs well beyond the region of electric field gradient, which is caused by distortion of an otherwise uniform field in the presence of the matrix. The concentrating effect exerted by each newly captured particle creates a new capture site. Thus the collecting capacity of the matrix of the same HGES device is much higher for metal ($\varepsilon \to \infty$) than for low dielectric constant particulates. F_e, acting on a dielectric in the vicinity of a matrix element, may exceed gravity by a factor of 50 to 150, and the capture cross-section in an electric field may exceed the filament radius by a factor of 50. The performance of a dielectrophoretic separator depends on the capacity of the matrix for selective capture and retention of the particles.

Typically the fiber diameter, in HGES unit which contains a mesh of very fine ceramic wool, is of the order of 25 μ , the electric field applied to the air gap by the electrodes is in the range of 10 to 25 KV, and the field gradient near the surface of the fibers (flux-converging elements) may be as high as 1,5 KV/μ for $BaTiO_3$ fiber. With such a device sufficient force will be applied to most materials to permit their filtration and separation.

---

* The porosity of the matrix in a WHIES device under different packing arrangements of uniform-sized spherical beads will be studied with a view to determining the influence of its variation on the free volume available for collecting particles in the "electrified working space".

The results of studies of regular geometric arrays of spherical beads can be used to evaluate purely mechanical trapping (tests at E = 0) vs combined dielectrophoretic-plus-mechanical trapping (tests at $E \neq 0$).

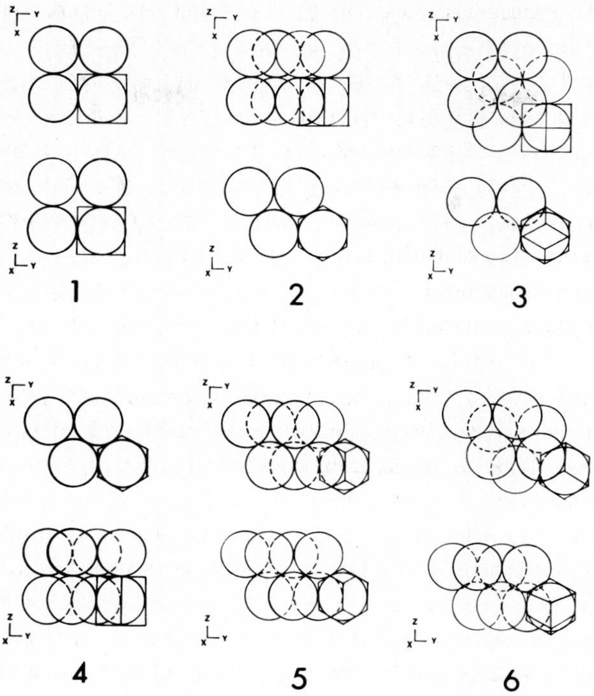

Fig. 22. Orthographic projections of aggregates of spheres for six stable packing arrangements. In each case, the top diagram shows the projection perpendicular to Z, and the bottom diagram shows the projection perpendicular to X. The unit polyhedron is shown in each case.

TECHNICAL REQUIREMENTS

It now seems appropriate to consider the properties which enable a dielectric separator to operate at a low cost per processed unit:

- high duty cycle
- high production rate
- high-quality S-S or S-L separation
- separation of particles even in the submicron range, including colored impurities
- simple automatic operation
- low loss of the pressure drop in the matrix
- simple maintenance
- low power requirement
- low capital and running costs.

Fig. 23 shows a schematic diagram of the cyclic HGES system which serves as laboratory set-up for filtration of non-conducting liquids. The basic components of the unit are a large volume of fine filamentary matrix, for which a variety of fiber and screen materials are available. On application of a strong electric field the matrix filaments develop extremely large trapping forces along their edges. The canister and matrix, which essentially constitute a filter bed, are placed between two electrodes. The fluid stream carrying the particles to be separated flows down through the canister with valves 1, 4 and 5 open with the electric field activated. Particles with $\varepsilon_p >_f$ are trapped by the matrix while those with $\varepsilon_p < \varepsilon_f$ pass unimpeded. Since the matrices used in the HGES device have a high void volume, usually about 95%, relatively low pressure drops are involved. When the matric has become saturated, the electric field is switched off and the particles are readily flushed from the matrix by closing valves 1 and 5 and opening valves 2 and 3.

Table 9 shows the conditions necessary for phase separation, either fractionation of various multi-component and poly-disperse systems, or fine purification of liquids by the removal of dispersed particles. Any technique for separating solids from liquid is based on application of a separating force to the solid particle so as to overcome the forces holding them in the liquid.

Separation efficiency and the duration of the separation cycle for any given S - L or S - S system is predetermined by means of laboratory-scale-tests. This ensures faultless performance of the system and a high collection rate and permits the process to terminate before oversaturation can set in. It also enables the optimum flow rate and operating voltage to be assessed.

When sharp separation is not attainable in a single pass, multiple treatment cycles are called for.

Capacity of cyclic batch units has so far been too low for suspensions where a large component is to be retained. In this case the devices fill and bridge too fast, and take too much to cycle for economic use. The approaches to the problem included shortening the cycle time and going to a continuously renewable filter matrix.

In principle, the two types of HGES and WHIES devices, the cyclic and the continuous, are similar to their HGMS and WHIMS counterparts. In the two first-named devices, the matrix may be static in the cyclic, and moving in the continuous variety. In a special design of a continuous unit, the matrix may be static and the pair of electrodes may move continuously from the feed station. The electric field is removed during flushing in a cyclic device, but retained in a continuous one.

Despite the simplicity of the carousel principle, the operation of continuous high-intensity unit is not entirely without problems: (i) scalping of very coarse solid fraction is necessary to avoid blockage of the matrix by coarse

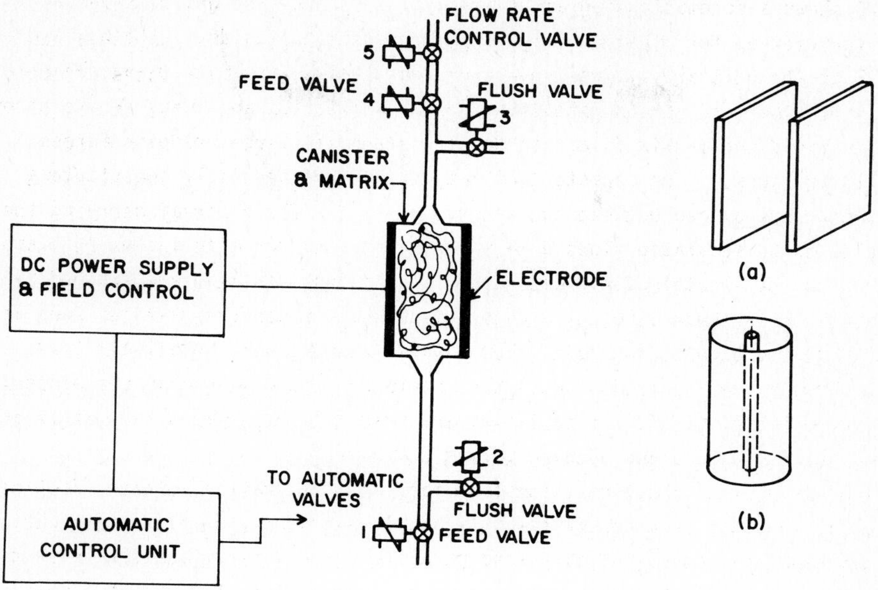

Fig. 23. Cyclic HGES unit (left); electrode configuration (right) (a) rectangular side by side, (b) cylindrical side by side.

TABLE 9

Conditions for Phase Separation by HGES or DF

	Phase Separation	ε Conditions
Gas purification	S - G	$\varepsilon_S > \varepsilon_G$
Fine purification of liquids	S - L	$\varepsilon_S > \varepsilon_L$
Fractionation of a binary mixture	S - S	$\varepsilon_{S_2} > \varepsilon_L > \varepsilon_{S_1}$
Separation of liquid mixtures	L - L	$\varepsilon_{L_1} > \varepsilon_{L_2}$

particles. (ii) for very short retention times the dielectrophoretic technique will not recover very fine particles (below 1μ) (iii) the selection of the matrix for a practical HGES or DF unit is governed by two factor; its collection efficiency, and the ease with which it can be cleaned in operation. A fine matrix will promote the high field gradients essential for separation and give better recoveries than a coarse one. However, it will be difficult to wash, and its dimensions will limit the top size of the feed.

Dielectrophoresis causes motion of the electrically neutral but polarizable particles, the direction of motion being independent of that of the field; in other words, either d.c. or a.c. (oscillating or pulsating) voltage can be employed, allowing for their different effect on the efficiency of S - S and S - L separation. As for the electric force, it varies as the field strength squared, hence it is independent of the sign of the field, so that it may be assumed that the effect of a.c. and d.c. voltage is the same in this case. From the mechanical viewpoint, a.c. voltage permits significant improvements in selective separation, by virtue of its jigging and shaking effect.

From an economic point of view, it is normally more feasible to produce steep electric gradients than to produce very intense electric fields. However, the modern HGES produces both steep gradients and intense electric fields, hence its effectiveness in separating materials having low dielectric constants. The high dielectrophoretic force that can be produced by specially designed matrices, without a power penalty, allows the slurry to be processed at a high velocity.

Dielectric seeding; particulate materials that would not be removed by HGES, WHIES, or DF can be seeded with conducting or high ε particles and then can be extracted. The seeding materials can be reclaimed by passing the filter washings through a dispersion stage followed by dielectric separation.
Dielectric scalping: for suspensions containing a high ratio of conducting to non-conducting particulates, it is best to scalp out the conducting material to avoid "bridging" the matrix and shorting out the high voltage field. When bridging occurs, restriction in flow ensues, which can lead to a complete stoppage of the stream. Scalping is usually accomplished by allowing a low intensity field to act on the pulp.

Cyclic dielectric filters are operated in the same manner as mechanical backflush filters. Prior to backflushing, the feed stream is diverted from the dielectric filter and the electrodes are de-energized. A short duration high velocity fluid (non conducting liquid, dry-compressed air, etc.) flush is used to purge the retained particles from the filter matrix.

A properly designed DF or HGES system requires that the retention of particles occurs throughout the entire depth of the unit (similar to deep bed filtration).

Dry separation: fluid viscosity of the liquid through the matrix plays an important part in determining the retention time necessary in the separator and in the limit of ε_p needed for a particle to remain trapped in the matrix as the fluid flows through it. The fluid also may add a cost of drying that would be desirable to avoid. For these reasons, dry separation should be tried in S - S separation processes. To date, all commercial use of HGES and DF has been in wet processing. However the technique is not inherently restricted to wet treatment. In dry as well as wet separation, the guiding principle in HGES and DF is the control of retention time.

Greatly improved collection efficiency as compared with conventional filters may as a rule be expected from the unique performance of DF when used in S - L systems whenever solid components are very finely divided at relatively low concentrations. Dielectrophoretic forces can be considerably larger than the viscous, inertial, or diffusion forces normally associated with the filtration process. The matrix is more permeable, hence allows higher filtration rates. These features indicate that DF is bound to play an important role in future filtration processes.

Some other features are: DF is non-plugging and exhibits a low pressure drop during separation; the system can be constructed in modular form, and operates continuously and automatically. Another major advantage is that the main operational parts of these devices are either stationary or moving slowly and hence the HGES process is highly reliable, requiring only a modest amount of maintenance.

In general, it can safely be stated that DF is an attractive alternative to conventional filtration/separation processes, such as cartridge filters, hydrocyclones, and sedimentation tanks, and it has, in fact, already been applied industrially.

WHIES, HGES and DF appear to be on the eve of a breakthrough, provided a sizable commercial development and marketing effort is initiated. Both the cyclic and the carousel type will contribute in commercializing this technology. Naturally, there are still some practical problems of hydraulic design flow distribution, controls, abrasion, flocculation, scaling, etc. to be solved for specific cases. To give an example: present WHIES devices physically plug with materials when handling particles $>150\,\mu$. This top size limitation is not set by the electric force constraints but by difficult slurry handling and matrix problems. The design of dielectrophoretic separators to handle large particle sizes is an area where research might well be justified.

MAIN APPLICATIONS

The only commercial applications of this technology to-date, are in petroleum refineries and the metal and vegetable-oil industries, where contamination by solids is a major problem in feedstocks, distillates, hydrocarbon-based products, etc. However, because of its ability to process colloid systems more rapidly than other more traditional separation technologies (barrier and depth filtration, centrifuging, settling/decanting, etc.), a wide range of applications in mineral processing, the petrochemical industry, waste treatment, and chemical and biochemical engineering, can be suggested and have actually been the subjects of experimental studies at the Mineral Engineering Research Center, Technion, under the National R & D Council of Israel sponsorship.

A prototype of a continuous Carousel (moving matrix) separator (see Fig. 24) was constructed at the Mineral Engineering Research Center on the smallest practical scale for use in laboratory tests on S - S and S - L separation. The

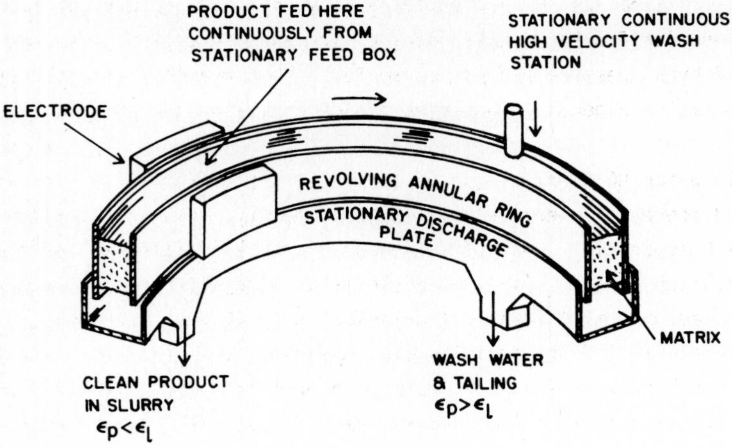

Fig. 24. Essential features of Carousel-type continuous WHIES unit - simplified sketch.

matrix moves continuously in and out of the electric field region in a wheel-like structure. The feed slurry enters the matrix while it is in the field region. Particles with $\varepsilon_p < \varepsilon_\ell$ pass directly into an annular launder below. The washing region is field-free, to allow release of the trapped particles. As the Carousel uses a moving matrix it is possible to reduce the dead time* and this means that a continuous unit has a greatly increased processing rate (see Equation 36).

As an illustration, the operation procedure began with the addition of 1 kg of MgO slurried in about 100 liters of light oil in a feed sump. The suspension was continuously stirred. A valve on the feed sump was adjusted to control flow rate and the valves of the Carousel were adjusted to balance the hydraulic heads. The field was switched on and the matrix set in rotation (0,01 to 1 r.p.m.). The electric field value for the tests was adjusted between 5 and 15 KV. Recoveries in the range of 80 - 95% were obtainable in a single pass.

---

*Fraction of the total time that the separator is not actually working.

Conventional filtration has proved to be very expensive in many applications due to the need for frequent replacement of the cartridges and to the combined conditions of high pressure and low capacity. Dielectric filtration promises minimized operating expenses, increased efficiency, and reduced loss of production time.

The ways in which HGES, WHIES and DF can be utilized are: (i) A mineral and a liquid with differential permittivity can be separated for clarification and polishing (S-L separation), or extraction of valuable constituents or removal of deleterious substances. (ii) Two dielectrics with differential permittivity can be separated even if they are of colloidal size (S-S separation). (iii) Materials with low permittivity can be separated by scavenging or coagulating with special seeding materials with high permittivity $\varepsilon_p > \varepsilon_f$ (such as colloidal TiO_2, titanates, metal powders, etc.) which can then be removed on the HGES unit. This modification may be called "piggy-back dielectric filtration". (iv) Separation is also possible when an organic dissolved species is capable of precipitation or co-precipitation onto a dielectric "carrier".

Diverse potential applications of HGES and DF may now be suggested (Ref.3,13).

In the metal-working industry

Removal of metallic fines from metalworking fluids (Ref. 10,35), e.g. in foil rolling, where contamination of the recirculated rolling oils makes for excessive pinholing. Also, removal of metallic debris from lubricating oils to reduce engine wear.

In the petroleum industry

Removal of fine particulates from refinery streams (Ref. 10,36), e.g. of FCC catalyst from decanted oil for production of low-ash carbon-black feedstock of fuel oil (see Fig. 25). High-volume feed streams for fixed bed reactors, such as hydrodesulphurizers, hydrotreaters, hydrofinishers, etc. contain low levels of solids, but in time these accumulate and plug the catalyst bed, causing a reduction in throughput and possible reactor shut-down.

In the chemical industry

For use with distillate streams, where suspended solids cause plugging in fixed-bed catalytic reactors. Particulate materials in these streams, such as coke particles, catalyst fines, and corrosion products, are easily separated. Recovery of weakly dielectric fine precipitates by deposition on highly dielectric particles may also be included under this heading.

Refinement of mineral concentrates

Ilmenite (Ref. 1, 37), cassiterite, titanium dioxide, etc., are amenable to

GULFTRONIC SEPARATOR SYSTEM
PERFORMANCE DATA
FCC DECANTED OIL

STREAM	TEMP °F	ASH CONTENT (WT %) BEFORE	AFTER
1	280	0.22	0.04
2	280	0.18	0.04
3	280	0.29	0.05
4	305	0.11	0.02
5	300	0.13	0.01
6	305	0.10	0.01
7	310	0.32	0.01

TYPICAL GULFTRONIC SEPERATOR INSTALLATION

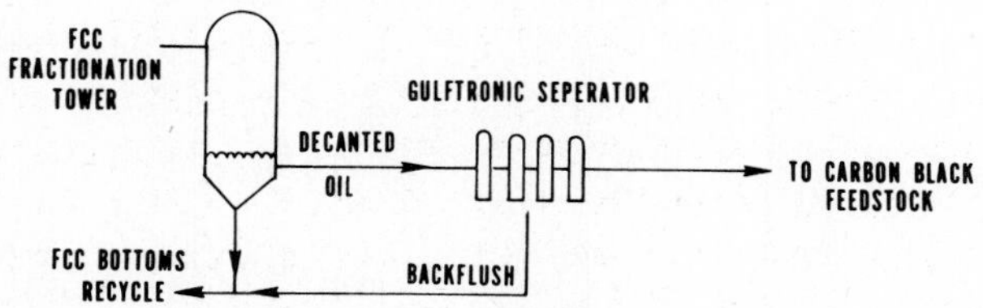

Fig. 25. The industrial Gulftronic batch separation system*: performance data (removing FCC catalyst fines from decanted oil).

*General Atomic Comp., P.O. Drawer 2038, Pittsburg, Pa. 15230, U.S.A.

the method, which is economical provided the price of the liquid is not prohibitive. Upgrading of precious metal concentrates (production of material of very high value from a low-volume starting material) is another example.

Removal of ash from pulverized, solvent-refined and liquefied coal

The liquefaction and gasification of coal results in a fantastic new batch of complex solid-liquid separation problems. Continuous mechanical filtration under high pressures and temperatures can be replaced by DF. Also, HGES can be applied to the problem of cleaning coal by direct mineral (ash) removal from coal slurries in oil. A potentially very wide field of application.

In the waste-reclamation industry

1. Recovery of industrial diamonds.

2. Treatment of effluents and waste liquids for recovery of dielectric materials. DF can be used to recondition petroleum products contaminated by debilitating particulates. Examples: used hydraulic fluids, liquid fuels, motor oils, and used transformer oils (Ref. 10 , 35, 38). The method applied does not change the original properties (viscosity, b.p., acidity, corrosion stability, etc.) of the fluid.

3. Regeneration of commercial frying oils which have picked up color and a significant solids content.

In processing of pharmaceuticals and cosmetics

Biological applications

An important point in dealing with biological dielectrophoresis is the strong dielectrophoresis is the strong dependence of the collectability of cells on the matrices with the frequency of the field and with the conductivity of the medium. Living cells in aqueous media can be moved, collected, and even separated by the dielectrophoretic technique (Ref. 39 - 41). Both DF and DL has been used to sort mixture of living cells (Ref. 42, 43). Microdielectrophoretic methods of analysing and separating biological particles, are also used to differentiate between normal and diseased cells (Ref. 14).

In fluid treatment

Suspended or dissolved dielectric solids are amenable to treatment by DF. Separation of low-dielectric suspended particulates can be achieved by means of "dielectric seeding" or "dielectric flocculation".

In the food industry

Examples of processing of edible fats and oils are dielectric filtration of nickel or copper hydrogenation catalysts or of decolorizing agents such as bleaching clays or activated carbon (Ref. 3), (See Fig. 26).

Accelerated sedimentation

Via dielectric precipitation of dispersed particles (Ref. 14, 44, 46).

L - L separation

To date, little work has been done in using HGES to separate liquid mixtures or to break emulsions. In this case separation is feasible if $\Delta\varepsilon = \varepsilon_{L_1} - \varepsilon_{L_2} > 0$. If $\Delta\varepsilon \rightarrow 0$, separation is effected by altering the dielectric properties of one of the fluids, e.g. selective admixture of titanates or metal powder followed by dielectrophoretic removal.

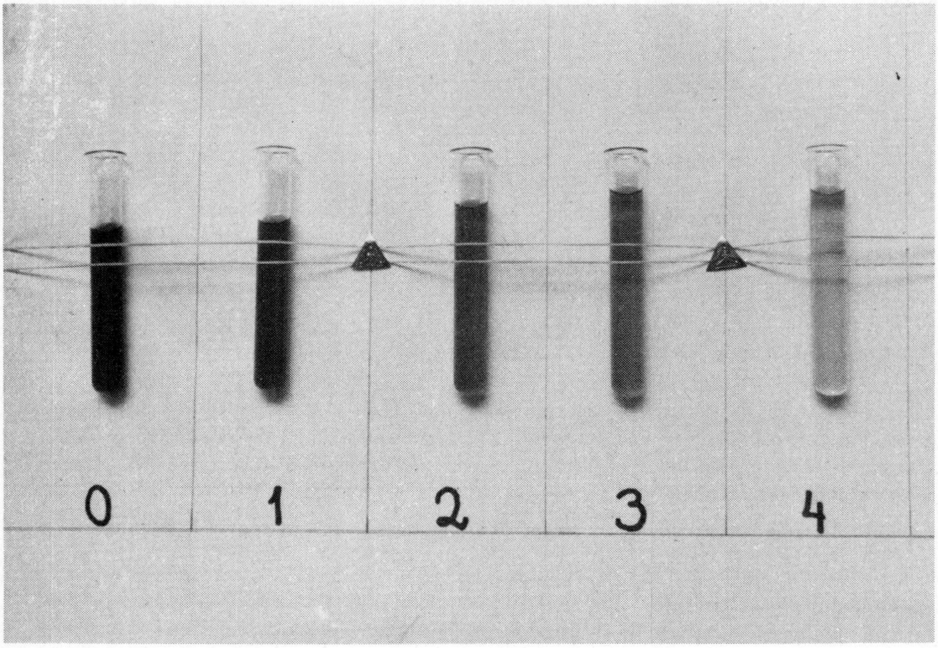

Fig. 26. Change in color of vegetable oil vs number of flow (Ref. 45) - improved brilliance and clarity.

S - G Separation

Mainly for air filtration (Ref. 47). HGES, WHIES, and DF may also serve as analytical or control tools for laboratory study: (i) Controlling the purity of non-conducting liquids, e.g. lubricating systems. (ii) Determination of wear and corrosion in liquids. The debris collected by this method are an important source of information as to the condition of an engine (abrasion and wear). (iii) Sensitive separation of monofractions for rapid testing and identification of minerals and other materials, e.g. for standard routines in mineralogical studies. (iv) Calculation of dielectric characteristics of dispersed systems.

Despite rapid commercialization of WHIES in the petroleum industry and some experimental studies in other areas (as mentioned above) in the last few years, HGES, WHIES and DF are only beginning to gain recognition by the technical community at large as a viable technology of significant potential impact.

Finally, we would like to repeat one of Hopstock's major findings and recommendations in reviewing magnetic and electrostatic separations in 1976 (Ref. 48): There are too few people and too few projects in dielectric separation research to generate enough ideas that will lead to major advances of the technology in a reasonable time.

DISCUSSIONS AND CONCLUSIONS

A high tension, high-gradient electrical analogue of the matrix-type magnetic separator (HGMS) and the dielectric filter was developed, tested and established.

The fundamental concept of dielectrophoretic separation is interaction of small conducting or non-conducting particles with specially designed matrices in an applied uniform background electric field. Each matrix element induces regions of highly non-uniform field intensity, which results in a net force being exerted on the particle. The dielectrophoretic force, in competition with viscous-, inertial-, and gravitational forces, causes $\varepsilon_p > \varepsilon_f$ particles to migrate to the surface of the matrix where they are retained until the dielectrophoretic force is removed.

Dielectric filtration is a subject of high diversity and of considerable interest and value to many industries. The potentialities of this unique technology are not sufficiently widely appreciated, and deserve much wider application. High-gradient electric separation is a new technique which provides a practical means for separating and fractionating dielectric materials down to colloidal particle size. This method operates on a large scale and at flow rates hundreds of times faster than those possible in ordinary mechanical filtration. It offers singular possibilities and a number of separations and demonstrable in the laboratory with minimal equipment.

The ability of non-uniform electric fields to produce separation in suspensions of conducting and non-conducting materials in non-polar fluid media is well established. Particle sizes down to the submicron range are amenable to this type of separation. The sensitivity of the separator's performance to changes in particle size, matrices geometry (size, depth, filling factor), flow rate, electric field strength, retention time, fluid temperature and particle dielectric properties may be predicted from simple separation tests on lab units.

It is clear that if HGES or WHIES devices are to be used on an industrial scale then continuous rather than batch operation is called for. The unique moving matrix separator, first developed at our laboratory, illustrates the essential features of Carousel-type continuous HGES and WHIES units. In addition, both cyclic and continuous HGES devices are amenable to automation and they can be quite easily fully integrated in the flowsheets of most modern industrial plants (Ref. 49).

With increasing particle fineness, the number of separation techniques available decreases rapidly. The ability of WHIES and HGES to process fines has led to many applications. Industrial applications so far have employed batch devices.

As the sizes of particles undergoing separation treatments become smaller, the surface forces gain in prominence and must be taken adequately into account. Consequently, a more intimate involvement of colloidal and surface chemistry with HGES process is desirable and likely in the future.

Several industrial processes were defined as potential candidates for the application of DF, WHIES and HGES. They could be a competitive technique to conventional methods. HGES potentially offers many important advantages over alternate methods; high collection efficiency for submicron particulates and high concentration efficiency on process performance.

The use of less viscous media - like air - will allow recovery at lower dielectric intensity and smaller particle size. The design of dry dielectric separators, designed to match the fields and gradients of the HGES, is an area for further research and development.

Also, DF, WHIES, and HGES are all well adapted to analytical laboratory work for mineralogical studies and ore dressing tests (Ref. 11). They serve in a variety of applications, such as dielectric material fractionation, dielectric filtration, static levitation, and permittivity measurement, and as a method of determining the electrical characteristics of particulates. They may also be used for pilot plant demonstration, simulating the operation of larger units, and for small-scale production.

We believe that lack of interdisciplinary communication is the main factor which has so far delayed other commercial applications. As new applications are conceived, continued research will undoubtedly be required for WHIES, HGES and DF to reach their full potential. Better understanding of the fundamental mechanisms of the processes are certainly needed. Electrode design, matrix size and configuration, and matrix cleaning methods must be tailored to specific applications for different particulate systems.

Experiment is still well ahead of the theory in many respects. Rapid progress will depend upon an hand-in-hand development of both the theory and experiments. In addition, HGES is a highly interdisciplinary technique and thus its progress can be expected to occur at the interface between applied electricity and other disciplines, such as: the theories of porous media, fluid dynamics, transport processes, electricity, surface sciences and colloid chemistry.

Summing up, it appears that HGES, WHIES and DF will almost certainly become an important technology in the next decade, penetrating the mineral, chemical, organic, biochemical, pharmaceutical and ink industries, being applicable to S - S, S - L or S - G separations and to purification of particulate systems.

These new separators have filled a need, caught the imagination of many and become a technology worth fighting for.

ACKNOWLEDGEMENT

The authors wish to thank the Mineral Engineering Research Center (MERC) and the Solid State Institute of the Technion - Israel Institute of Technology, for providing laboratory space and technical assistance. Financial support of this work by the National Research and Development Council of Israel (grant 016-098) is also gratefully acknowledged.

REFERENCES

1. I.J. Lin, I. Yaniv and Y. Zimmels, On the separation of minerals in high-gradient electric field, XIII Inter. Miner. Process. Cong. (IMPC), Warsaw, Poland, June 4-9, Vol. II (1979), 83-105.
2. Y. Zimmels, I.J. Lin and I. Yaniv, Advances in application of magnetic and electrical techniques for the separation of fine particles, Proc. AIME Inter. Symp. on Fine Particles Processing, Las Vegas, Nevada, U.S.A., Feb. 24-28 (1980), Vo. II, 1155-1177.
3. I.J. Lin, L. Benguigui, Sep. & Purification Methods, 10, No. 1 (1981), 53-72.
4. I. Yaniv, Y. Zimmels and I.J. Lin, Sep. Sci. & Tech., 14, No. 4 (1979), 261-290; I. Yaniv, I.J. Lin and Y. Zimmels, IEEE Trans. on Mag., 14 No. 6 (1978), 1175-1179; Y. Zimmels and I.J. Lin, Ibid, 18 No. 3 (1982), 921-928.
5. E. Cohen, IEEE Trans. Mag., 12 No. 5 (1976), 503-506.
6. F. Fraas and O.C. Ralston, Electrostatic separations of solids, Ind. Eng. Chem., 32 (1940), 600-604.
7. I.J. Lin and L. Benguigui, Powder Tech., 17 No. 1 (1977), 95-100; L. Benguigui and I.J. Lin, J. Appl. Phys., 49 No. 4 (1978), 2536-2539; I. Yaniv, I.J. Lin and Y. Zimmels, Sep. Sci & Tech. 14 No. 7 (1979) 557-570.
8. T.B. Jones and G.A. Kallio, J. of Electrostatics, 6 (1979), 207-224; T.B. Jones and G.W. Bliss, J. Appl. Phys., 48 (1977), 1419; A. Watanabe, Japan. J. Appl. Phys., 14 No. 9 (1975), 1301-1306; D.S. Parmar and A.K. Jalaluddin, Japan. J. Appl. Phys., 13 (1974), 793.
9. H.A. Pohl and K. Pollock, J. Electrostatics, 5 (1978), 337-342; C.M. Feeley and H.A. Pohl, J. Phys. D. 14 No. 11 (1981), 2129-2138.
10. G.R. Fritsche, The Oil & Gas J., 75 No. 13 (1977), 73-77; U.S. Patent No.3, 928, 158, Dec. 23 (1975); S. Africa Patent Appl. No. 81/0941, Feb. 12 (1981).
11. I.J. Lin and L. Benguigui, Sep. Sci. & Tech., 17 No. 5 (1982), 645-654; L. Benguigui and I.J. Lin, Ibid, 17 No. 8 (1982), 1003-1017.
12. L. Benguigui and I.J. Lin. J. Appl. Phys., 53 No. 2 (1982), 1141-1143.
13. I.J. Lin and L. Benguigui, 3rd World Filtration Congr. 13-17 Sept. (1982) Philadelphia, U.S.A.
14. H.A. Pohl, "Dielectrophoresis: The behaviour of neutral matter in non-uniform electric fields", Cambridge Univ. Press, Cambridge, 1978, 579 pp.
15. C.E. Jordan, et al., A continuous dielectric separator for mineral beneficiation, USBM, RI 8437 (1980), 18 pp.
16. F.J. Friedlaender, M. Takayasu and W.H. McNeese, IEEE Trans. on Mag., 15 No.6 (1979), 1526-1528.
17. R. Birss and M.R. Parker, Ibid, 15, No. 6 (1979), 1523-1525.
18. W.H. McNeese and P.C. Wankat, Ibid, 16, No. 5 (1980), 840-842.
19. I.I. Kalyatskii, et al. Comparative study of the mechanical and electrophysicalgrinding, Redk. Elem., 9 (1973), 115-125.
20. E. Sarapuu, Electro-energetic rock breaking systems, Min. Congr. J., 59 No.6 (1973), 44-54.
21. A.T. Bondarenko, Study of the temperature dependence of the dielectric constant of rocks at different frequencies, Bull. Acad. Sci. (USSR), Geophys. Ser., No. 3 (1963), 281-286.
22. E.I. Parkhomenko, Electrical properties of rocks, Plenum Press, N.Y. (1967), Transl. from Russian, ed. by G.V. Keller, 314 pp.

23. A. von Hippel, (ed.), Dielectric materials and applications, Technology Press of MIT and John Wiley & Sons, Inc., N.Y. (1954), 438 pp.
24. A.H. von Hippel, Dielectrics and Waves, John Wiley & Sons, Inc., N.Y. (1954), p. 39.
25. Landau and E.M. Lifshitz, Electrodynamics of continuous media, Pergamon Press, Lond. (1960).
26. J.W. Rayleigh, Phil. Mag. 34 No. 5, 481, (1892); D. Bergman, Phys. Rev., 14B, 4303-12 (1976)
27. C.J.F. Böttcher, Rec. Trav. Chim., 64 No. 47 (1945); C.J.F. Böttcher, Theory of Electric Polarization, Elsevier, Amsterdam (1952) Ch. 9.
28. J.C. Maxwell-Garnet, Phil. Trans. R. Soc. Lond., 203, Nos. 3 & 5 (1906).
29. I. Webman, J. Jortner and M.H. Cohen, Phys. Rev., 15B 5712-5723 (1977).
30. G. Molinari and A. Viviani, J. Eelectrost. 5 (1978), 343-354.
31. B. Jaffe, W.R. Cooke and H. Jaffe, Piezoelectric Ceramics, Academic Press, New York (1971).
32. J.J. Martin et al., Pressure drop through stacked spheres: effect of orientation, Chem. Eng. Prog. 47 No. 2 (1951), 91.
33. N. Ouchiyama and T. Tanaka, Ind. Eng. Chem. Fundam., 19 (1980), 338-340.
34. G. Zebel, J. Colloid Sci., 20 (1965), 522-543.
35. Filtration and Sep., 14, No. 2 (1977), 140-144.
36. A.A. Gundyrev, L.P. Kazakova and Z.Y. Oleinik., Chem. Tech. Fuels Oils, 12, Nos.7-8 (1976), 595-598.
37. H.A. Pohl and C.E. Plymate, J. Electrochem. Soc., 107, No. 5 (1960), 390-396.
38. G.A. Nixitin, Electrical cleaner of dielectric liquids, Khim. Tekhnol. Topl. Masel, No. 4 (1980) 21-24 (C.A. 93:134192t).
39. H.A. Pohl and I. Howk, Science, 152 (1966), 647.
40. H.A. Pohl, J. Biol. Phys., 1, No. 1 (1973), 1-16.
41. H.A. Pohl, K. Kaler and K. Pollock, J. Biol. Phys., 9 (1981), 67-86.
42. L.E. Cross and L.H. Hardy, Ferroelectrics, 10, (1976), 241.
43. K. Kaler and H.A. Pohl, J. Biol. Phys., 8, (1980), 18-31.
44. H.A. Pohl, Sci. Am., 203 No. 6, (1960), 107-116.
45. I.J. Lin and L. Benguigui, Dielectric Separation, Merc. Rep. 016-098, July (1981), 100 pp.
46. A.A. Gundyrev, L.P. Kazakova and Z.Y. Oleinik, Chem. Tech. Fuels oils, 12 No. 7-8 (1976), 595-598.
47. G.O. Nelson and C.P. Richards, Air filtration enhancement using electronic techniques, Proc. ERDA Air Clean Conf. 14th, Sun Valley, Idaho, Aug. 4, (1976).
48. D.M. Hopstock, Electrostatic and magnetic separations, Research needs in Mineral Processing; P. Somasundaran and D.W. Fuerstenau, eds. Columbia Univ. (1976), 88-97.
49. I.J. Lin and L. Benguigui, Coal, Gold and Base Minerals J., 30 No. 7 (1982), 45-65.

Progress in Filtration and Separation 3, blz. 205-266

THE DRYING OF POROUS MATERIALS WITH ELECTROMAGNETIC ENERGY GENERATED AT RADIO AND MICROWAVE FREQUENCIES

R. M. PERKIN
ELECTRICITY COUNCIL RESEARCH CENTRE, CAPENHURST, CHESTER, CH1 6ES, ENGLAND

ABSTRACT

The principles of drying porous materials with electromagnetic energy generated at radio and microwave frequencies - dielectric heating - are reviewed. Established methods of power generation and application, the physical background to drying, and the drying characteristics derived from theoretical modelling are discussed. Industrial applications and present developments are considered.

CONTENTS

CONTENTS cont'd

1. INTRODUCTION

The removal of moisture from porous solids by evaporation is a common industrial operation. The desire for more efficient drying and improved product quality has resulted in many new techniques and types of dryer appearing in recent years to replace the older forms of drying. One such technique is drying with dielectric heating, that is the volumetric heating of a wet material by electromagnetic energy generated at radio or microwave frequencies. This method utilizes the ability of polar liquids, most commonly water, and liquids containing salts to absorb electromagnetic energy at these frequencies. In contrast to other heat transfer methods, the energy is transmitted directly to the volume of the wet material and does not rely on conduction of heat from the surface. This characteristic allows increased heat transfer which speeds up the drying process and offers other advantages over conventional methods.

The capital cost of radio frequency (r.f.) and microwave equipment is high and without doubt many applications, although technically feasible, are at present uneconomic. However, past experience has shown that where these methods are used the payback due to the improved process has far outweighed the initial extra cost. For this reason it seems worthwhile for the engineer to be acquainted with these r.f. and microwave drying techniques since at sometime they may provide the solution when a problem arises because traditional processes fail to cope with new constraints or when new production methods are sought.

In the present work the basic principles of electromagnetic energy generation and application are reviewed. The physical background to drying with these techniques is outlined and the drying characteristics as obtained from theoretical modelling are discussed and compared with experimental results.

The practical application of these methods and the advantages which they offer are then summarised and present developments are indicated.

2. ELECTROMAGNETIC POWER ABSORPTION

The subject of radio frequency and microwave heating is likely to be unfamiliar to the chemical or process engineer. For this reason the basic principles and established methods of radio frequency and microwave power generation and application will be reviewed. Further details on radio frequency methods can be found in references (1) and (2) while the subject of microwave heating is dealt with comprehensively in the recent book by Metaxas and Meredith (ref. 3). Developments in the use of microwave and to a lesser extent r.f. heating can be found in the Journal of Microwave Power.

2.1 General Considerations

The electromagnetic power absorbed in a unit volume of dielectric material and dissipated as heat is given by

$$q = 2\pi\varepsilon_o f \varepsilon'' E^2 \quad W/m^3 \tag{2.1}$$

where f is the frequency of the applied electromagnetic field, ε_o the permittivity of free space, ε'' the loss factor and E the electric field strength within the material.

To avoid interference with communication channels, dielectric heating equipment can only be operated at certain permitted frequencies. The most commonly used are: 13.56, 27.12 MHz for radio frequency and 896, 915 and 2450 MHz in the microwave region.

A comparison of the general features of the systems for the two frequency ranges is shown in table 2.1. The extent to which a material can absorb energy in these frequency ranges is represented by the magnitude of the loss factor of the material ε'', a dimensionless quantity determined by the electrical properties of the medium. The presence of moisture within a solid generally increases the value of the loss factor, as compared to that for the bone dry material, by several orders of magnitude. For wet material, ε'' is a function of moisture content, temperature and frequency.

The electric field strength within the material depends on the particular configuration of the electrical circuit which produces the electric field and the dielectric constant of the material ε. A general requirement is that the electric fields associated with the equipment must be below the ionizing

TABLE 2.1
General characteristics of radio frequency and microwave techniques

Characteristic	Radio frequency	Microwave
Frequency/wavelength (free space)	13.56 MHz/22.2m 27.12 MHz/11.1m	896 MHz/0.3m 2450 MHz/0.12m
Power source	Class 'C' triode	Magnetron
Generation efficiency	50 - 80%	70-80% 896 MHz 50-60% 2450 MHz
Average efficiency of power transfer from the mains supply to product	$\lesssim$ 50%	$\lesssim$ 60%
Power rating of single, commercially available unit.	600kW typical units 12,25,50,75kW.	30 - 60kW 896 MHz 6kW 2450 MHz
Principal mechanism for power absorption	Ionic conduction, smaller dipole component	Rotation of dipoles, ionic conduction
Waveguide dimensions	-	(0.25 x 0.125)m 896 MHz (76 x 38)mm 2450 MHz

breakdown value for the surrounding air. For clean planar electrodes at atmospheric pressure this breakdown field is equal to 3000 kV/m. As the pressure is reduced the breakdown field decreases reaching a minimum value $\sim$ 8 kV/m at a pressure around 100 Pa (1mbar abs). Microwaves generated at 2450 MHz have the largest breakdown field at these pressures while r.f. has the smallest (ref. 3).

At atmospheric pressures the maximum permissible electric field within the material is of the order of 100 kV/m. Using this value of electric field strength, the range of maximum power densities attainable in drying applications are presented in table 2.2 for two values of ε''.

TABLE 2.2
Maximum practicable power densities kWm^{-3} for E = 100 kV/m

Frequency (MHz)	ε" = 0.1	ε" = 1
13.5	75	$7.5\ 10^2$
27.12	$1.5\ 10^3$	$15\ 10^3$
896	$5\ 10^4$	$5\ 10^5$
2450	$1.36\ 10^5$	$1.36\ 10^6$

The lower value of ε" is for nearly bone dry solids and the upper for solids saturated with liquid or with salts present in the moisture.

2.2 Dielectric Properties

The most important term in equation (2.1) is the loss factor since its value decides whether dielectric heating methods are feasible, dictates which operating frequency must be used and sets an upper limit on the power density which can be achieved.

The effective loss factor for a wet body is derived from that of the skeleton solid, the bound liquid and the free liquid. For convenience, water will be taken as a typical case of a polar liquid. Provided the free water is present in sufficient quantity, it largely determines the effective value of the loss factor. As the solid dries out the contributions from the bound water and the skeleton solid, although small in absolute value can become important.

The loss factor of free water is given by

$$\varepsilon''_L = \frac{\sigma}{2\pi\varepsilon_0 f} + \varepsilon''_{DIPOLE} \qquad (2.2)$$

where the first term arises from the ionic conductivity σ due to any dissolved salts which may be present and the term ε''_{DIPOLE} represents the 'loss' due to the rotation of dipolar water molecules in the applied electric field. As Figure 1a shows, in the absence of dissolved salts ($\sigma = 0$) the maximum value of ε_L'' lies in the microwave region at about 17 GHz; at radio frequencies ε_L'' is insignificant by comparison. However, if sufficient amounts of dissolved salts are present the value of ε_L'' can be large at radio frequencies and the power absorbed in both frequency ranges can be appreciable, Figure 1b. Both σ and ε''_{DIPOLE} are temperature dependent: σ increases and ε''_{DIPOLE} decreases with increasing temperature (ref. 3).

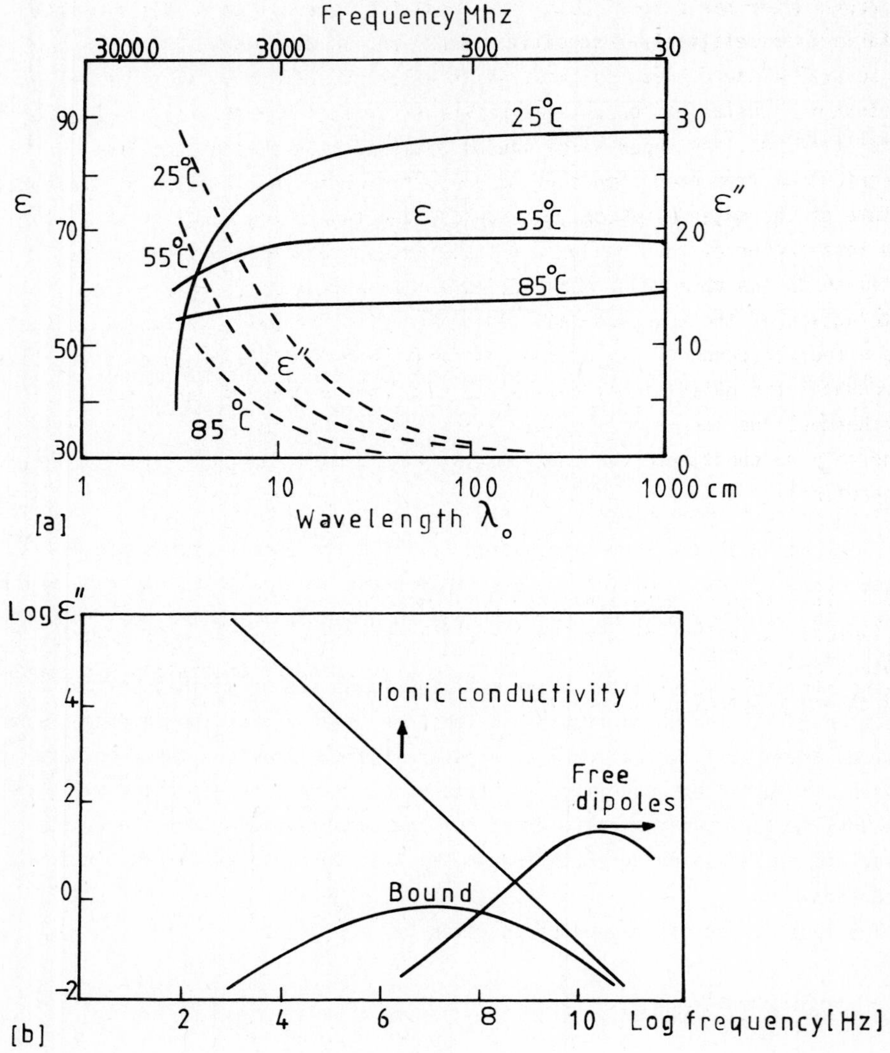

Fig. 1. Dielectric properties of water, (a) variation of dielectric constant ε and loss factor ε" for the free dipole component, (b) common constituents of ε_L''; → indicates the shift of the curve with increasing temperature.

Bound water has a loss factor intermediate between that of ice and free water and is relatively insensitive to changes in temperature.

Several mixture theories exist which allow the effective loss factor and dielectric constant to be estimated from the values for the solid and free water (ref. 4). An upper value can be obtained by assuming that the contribution from each component is proportional to the fraction of the total volume of the material which it occupies, and then using equation (2.2) for the loss factor of the free water. Such an approach is useful as a first estimate in the absence of more reliable data but it does neglect the interaction of the ions and water molecules with the solid and the fact that the effective conductivity of the ions when held in isolated threads of liquid throughout the material may not be the same as the measured d.c. value. Furthermore the loss factor can be anisotropic. For accurate results there is generally no substitute for experimentally determined values of the dielectric properties.

2.2.1 Variation with moisture content

Typical variations of loss factor and dielectric constant with moisture content are depicted in Figure 2 for a non-hygroscopic, moderately hygroscopic and strongly hygroscopic material.

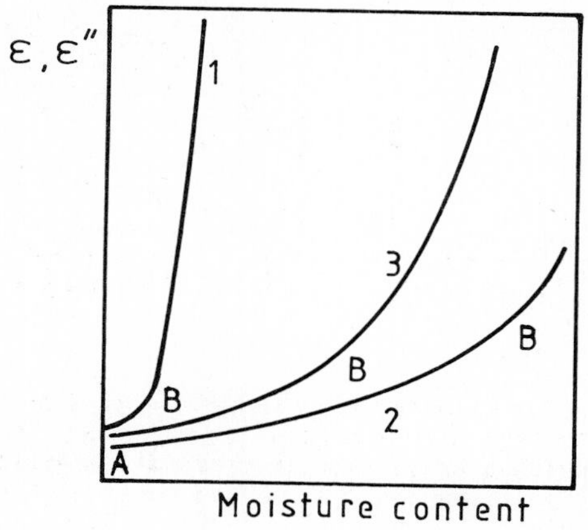

Fig. 2. General behaviour of the dielectric properties of a wet solid (1) non-hygroscopic (2) strongly hygroscopic (3) moderately hygroscopic. AB region of tightly bound moisture.

At low moisture contents when the water molecules are bound in a monolayer on the pore walls there is little energy absorption and the loss factor is close to that of the skeleton, region AB, Figure 2. As the moisture content X increases and the water forms multilayers its properties tend to those of the free, unbound liquid and ε'' and $\partial\varepsilon''/\partial X$ increase. The change in loss factor with moisture content can exhibit a sharp transition between the bound and the unbound regions (typical at radio frequencies) or may be relatively smooth so that the definition of a critical moisture content at which the transition occurs is somewhat arbitrary. Nevertheless, for strongly hygroscopic materials the upper end of the region AB lies in the moisture content range of 10 to 40% (dry basis) whereas for non-hygroscopic materials it lies at about 2%. The corresponding values for moderately hygroscopic materials lie in

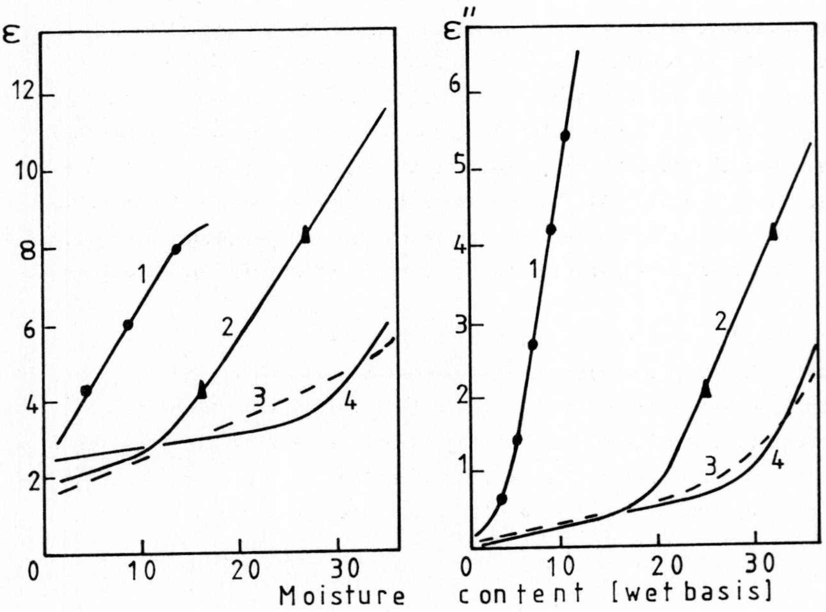

Fig. 3. Dielectric properties of some typical materials, measured at 2450 MHz (1) Sand (2) Polyamid (3) silica gel (4) mashed potato powder. After S.S. Stuchly, J. Microwave Power 62-68, 5(2), 1970; published by permission of the International Microwave Power Institute.

between. Typical variations of loss factor with moisture content are shown in Figure 3. The large increase in loss factor with moisture content for non-hygroscopic and moderately hygroscopic materials can be used to great effect to selectively heat the wettest regions of a solid and thereby reduce moisture profiles (see later sections).

2.2.2 Variation with temperature

For a non-hygroscopic solid as the temperature is raised the component of the loss factor due to the ionic conductivity increases and the dipole component decreases. At microwave frequencies the latter dominates so that for a given moisture content, $\partial\varepsilon''/\partial T$ is negative. In contrast, at radio frequencies the conductivity term is dominant and $\partial\varepsilon''/\partial T$ is positive.

With hygroscopic materials the relative proportion of bound liquid decreases as the temperature is increased so that for a given moisture content, more free, unbound liquid becomes available to absorb electromagnetic energy. This effect is most noticeable for weak or moderately hygroscopic materials where the temperatures required to liberate the bound moisture are not excessive ($\lesssim 100°C$). As a consequence the dipole component can exhibit an increase with increasing temperature if there is a large enough proportion of bound moisture initially present and the overall effect is for $\partial\varepsilon''/\partial T$ to be positive.

At room temperatures the loss factor of the skeleton solid is seldom significant. However some materials, for example nylon and acrylic, exhibit a rapid "runaway" increase in loss factor with temperature. Under these circumstances r.f. and microwave heating cannot be used if the required final moisture content is close to bone dry or if dried out areas may be present as a result of treatment prior to the dielectric heating stage.

2.2.3 Measurement techniques and available data

At radio frequencies the dielectric properties are measured using a tuned resonant circuit employing an inductance and capacitor. When the sample is placed between the capacitor plates the resonant frequency and the voltage across the capacitor change. The circuit is returned to the original frequency and voltage using variable capacitors and resistances. The loss factor and dielectric constant can then be related to the increments in capacitance and resistance (ref. 5).

Microwave measuring techniques utilise the change in wave propagation characteristics of a coaxial waveguide when material is inserted into the waveguide. From the standing wave patterns which are set up it is possible to deduce values of ε and ε'' (ref. 5).

For dry solids and liquids the above techniques are relatively straightforward. With wet solids, measurements become increasingly difficult as the temperature is raised. This is due to the problem of maintaining a humid atmosphere in equilibrium with the solid in order that the desired moisture contents can be obtained.

Although an appreciable amount of data exists for paper, coal, textile and food products, mainly at microwave frequencies and room temperature, there is a general lack of data for most other materials. An inexpensive, reliable, rapid means of measuring dielectric properties as a function of frequency, moisture content and temperature has still to be perfected. A review of microwave data and an extensive list of references can be found in references (3) and (6).

2.3 Power generation and application at radio frequencies

2.3.1 Generation

Radio frequency power for industrial systems is generated with a standard self-excited oscillator circuit operating under class 'C' conditions using an industrial triode valve with an air or water cooled anode (ref. 1).

The electrical equivalent of an r.f. dryer is illustrated in Figure 4. In

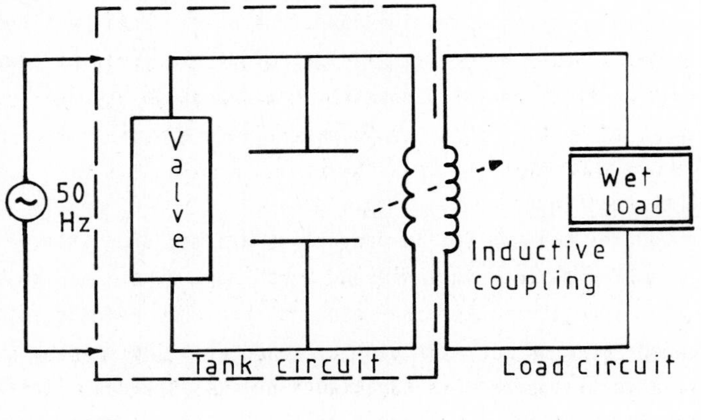

Fig.4. Electrical equivalent of a radio frequency dryer

operation the power is generated at a frequency determined by the capacitance and inductance of the tank circuit. This power is fed to the "load" or applicator circuit via inductive coupling and/or an electrical transmission line. The applicator circuit consists essentially of an inductance and a capacitor or "applicator" which induces the electric field in the material to be heated - the load. The combination of the generator and applicator circuit constitutes a resonant system and in order to generate power and transfer it efficiently to the load, the resonant frequency of the load circuit must be the same as that of the tank circuit and the inductive coupling between the circuits should be at some critical value determined by the valve characteristics and the dielectric properties of the load material. As the properties of the load change, as would be the case, for example, with a static load which was being dried, the system would detune and little or no power would be generated. This feature can be beneficial since it prevents overheating of the load. On the other hand it can mean that special coupling units may be required to keep the system on resonance and at the correct coupling in those circumstances where the load changes during its initial heating-up period (ref. 7). These problems of tuning r.f. systems are reduced by using a continuous rather than batch process to give quasi-steady load conditions.

At present the maximum output for a commercially available r.f. generator is around 600kW. The efficiency of conversion of mains electricity to r.f. power within the generator is typically between 55-70% and the efficiency of energy transfer from the generator to the load is $\lesssim$ 80% so that in terms of energy input from the mains to the product an efficiency $\lesssim$ 50% should be expected.

The extent to which power is absorbed by the wet material itself, rather than dissipated by ohmic heating of the metal components of the applicator circuit, is determined by the so-called "Q-factor" of the applicator when empty (unloaded) and when 'loaded' with the wet material. The Q-factor of an a.c. circuit is defined as

$$\text{Q-factor} = 2\pi \times \frac{\text{Peak energy stored in the circuit}}{\text{Energy dissipated in the circuit per cycle}}$$

A large unloaded Q-factor means that little energy is absorbed by the metal components while a small loaded Q-factor signifies that the wet material is capable of absorbing the electromagnetic energy. For high efficiencies the unloaded Q-factor should be large (400-1000) and the loaded Q-factor small (30-100).

2.3.2 Application

The physical shape of the applicator capacitors can be designed around the product to be dried subject to the conditions that the correct resonant frequency, suitable Q-factors and an acceptable electric field (E-field) distribution are obtained. The most common examples are shown in Figure 5.

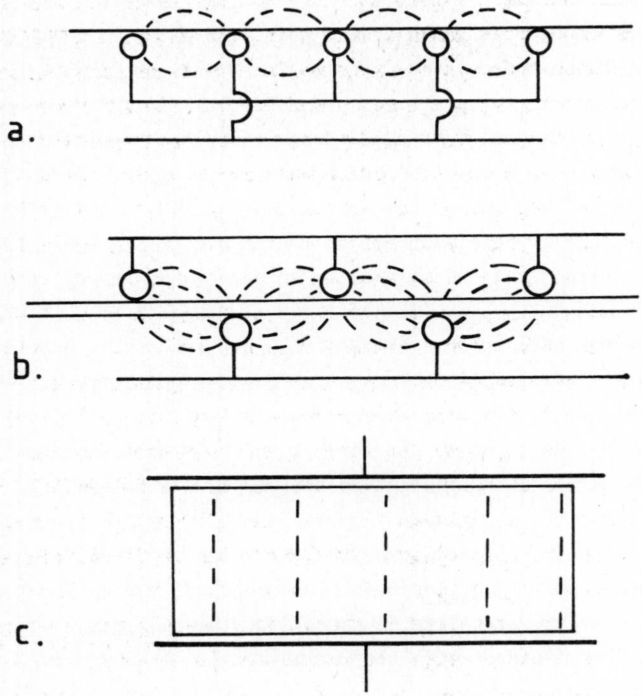

Fig. 5. Common radio frequency applicators
(a) Strayfield (b) Staggered throughfield (c) Parallel plate throughfield
--- Electric field distribution without load

The parallel arrangement of cylindrical electrodes where adjacent cylinders are of opposite polarity, known as the strayfield configuration, is only suitable for sheet material or thin beds. Improved loaded Q-factors and a more even field distribution are obtained when the relative positions of the electrodes are altered to give the staggered throughfield arrangement of Figure 5b which can handle thicker materials. For large objects, plate electrodes, the so-called throughfield, can be used, see Figure 5c.

The cross-sectional E-field distribution for the above applicators when loaded with material differs from the unloaded patterns shown in Figure 5. For the strayfield and staggered throughfield configurations the electric field within the material is bent so that the E-field is approximately parallel with the web or sheet of material. For these systems the change in magnitude of the electric field through the dryer is relatively small to the extent that the variation in the absorbed power, see equation (2.1), follows closely that of the loss factor. As seen in section 2.2.1 for weakly hygroscopic or unbound moisture the loss factor increases steadily with moisture content and consequently the power is preferentially absorbed in the wettest parts of the material. The value of this important feature will be examined later.

The presence of a dielectric load between the plates of the throughfield applicator does not alter the uniform E-field distribution. Furthermore, provided the inevitable air gap between the load material and the top electrode is small relative to the height of the load, the magnitude of the electric field remains constant even if the dielectric constant changes through the load. As with the other electrode systems the variation of absorbed power follows that of the loss factor. The air gap must be kept small to avoid the need for an excessive electric potential difference between the plates which could lead to arcing from surface imperfections on the high voltage electrode.

If the length of the electrodes is $\gtrsim 0.1.\lambda_0/\sqrt{\varepsilon}$ where λ_0 is the free space wavelength then variations in the E-field due to standing wave patterns can occur. For drying applications these can be largely overcome by suitably positioning the voltage leads on the applicator (ref. 2).

2.4 Power generation and application at microwave frequencies

2.4.1 Generation

Magnetrons and to a lesser extent klystrons are used to generate microwave energy in industrial systems. At the frequency of 896/915 MHz the usual operating output of a single unit is 30kW with a mains to microwave conversion

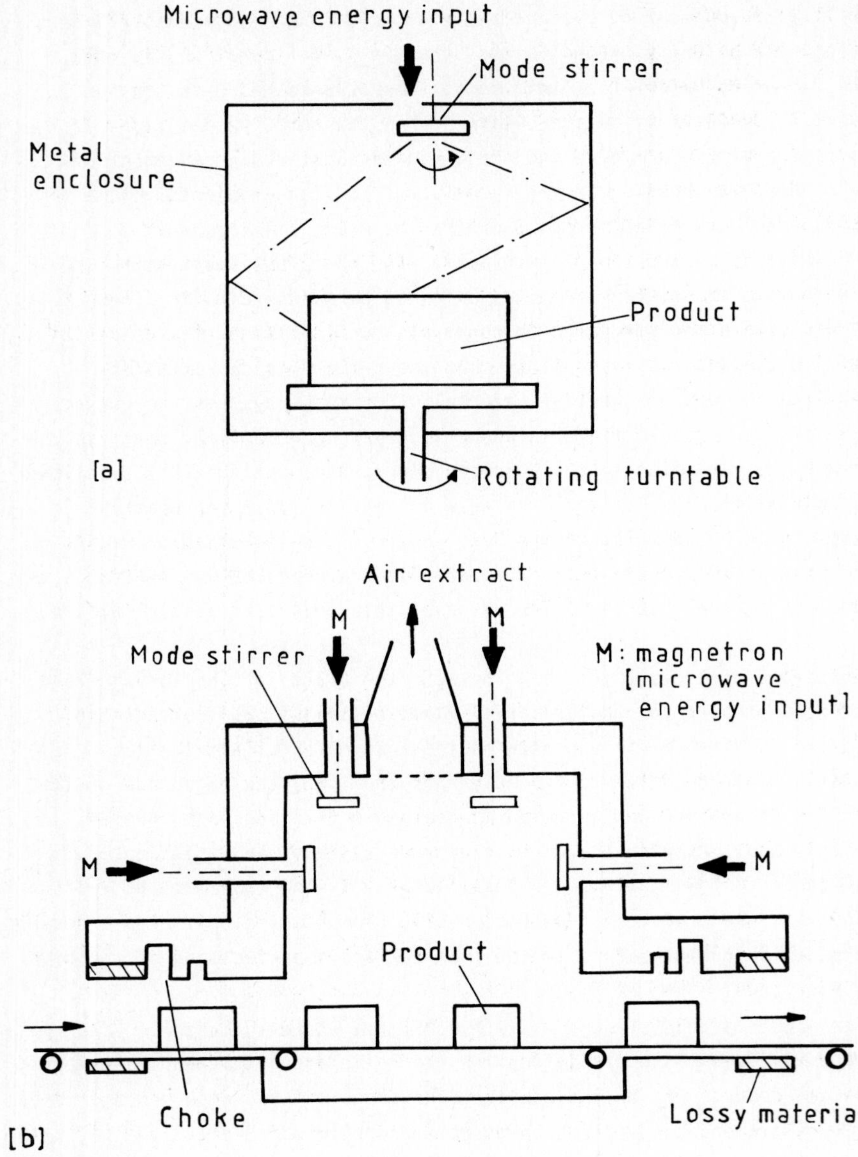

Fig. 6. Multimode microwave oven (a) for batch processes and (b) for continuous processes.

efficiency of 70-80%. The maximum available output for a single unit at 2450 MHz is 6kW with a conversion efficiency of about 50-60%. The power produced by the magnetron or klystron is transmitted to the microwave applicator by means of conducting metal waveguides. Unlike r.f. generators, the power is produced irrespective of the coupling or tuning with the applicator so some provision must be made to dissipate the energy generated in the event that it is not absorbed by the load, otherwise the magnetron can be damaged. Usually a dummy water load is incorporated and/or magnetrons which can withstand an appreciable amount of reflected energy are used. Energy reflected to the magnetron generally reduces the generation efficiency and shortens the lifetime of the magnetron and is to be avoided as far as possible.

2.4.2 Application

The applicators presently used are predominantly of three types:- the multimode oven, the travelling wave applicator and single mode resonant cavities.

Multimode oven

The multimode oven consists of a metal box or cavity, partially filled with the material to be heated, into which microwaves are injected. The electromagnetic waves are reflected from the walls of the cavity and standing waves with nodes and antinodes are set up, Figure 6a. In order to obtain an even E-field distribution the number of modes of oscillation of the electromagnetic field present in the cavity should be as large as possible and the antinodes should be close together. This is achieved by using the smaller wavelength at 2450 MHz, having the cavity dimensions as large as practicable, and launching the microwave energy through more than one port. Further improvements are obtained when a 'mode stirrer' in the form of a metal paddle is introduced to vary the standing wave pattern. Despite these measures some non-uniformity in the electric field, albeit small, is still to be expected. The introduction of the material into the cavity changes the field pattern and in general there is an E-field distribution within the material. This arises because when microwaves are incident on a dielectric the transmitted power is gradually absorbed and the magnitude of the E-field diminishes. For microwaves impinging on one face, the distance in which the transmitted power drops to 36.8% of its initial value at the surface is given by

$$\text{attenuation distance} \simeq \frac{\lambda_0 \sqrt{\varepsilon}}{2\pi \, \varepsilon''} \quad (\text{m}) \qquad (2.3)$$

For most practical cases this distance is only a fraction of the wavelength, usually 0.1 to 1. This attenuation of the electric field limits the thickness of the material or bed depth of particles which can be successfully heated. Whilst progress is being made on the theoretical prediction of the electric field within multimode ovens, (ref. 8), their actual design is still based on simplified models, physical considerations and previous practical experience.

The power transfer from the magnetron is relatively insensitive to the changing properties of the load down to low moisture contents and so multimode ovens are suitable for batch or continuous processing of materials in all but thin sheet form. Continuous systems must include filters at the product inlet and outlet ports to prevent leakage of radiation, Figure 6b.

Travelling wave applicator

The travelling wave applicator consists of a rectangular waveguide modified by cutting a slot in its side so that material in sheet form or held on a belt can be passed continuously through the waveguide at the position of maximum electric field, Figure 7. For a 25kW generator operating at 896 MHz the electric field strength in the waveguide is typically 20-50kV/m, well below the breakdown strength of air (ref. 9).

As the electromagnetic wave passes along the waveguide the energy is dissipated in the load and the electric field is attenuated. The power loss after traversing a section Δz of the waveguide can be expressed in terms of the input and output powers to the section by

$$10 \log_{10} (\text{Power in/Power out}) = 2\alpha\, \Delta z \quad \text{dB} \qquad (2.4)$$

where for a slab of thickness 2d, with constant dielectric properties, placed symmetrically in the waveguide the parameter α is given by (ref. 3)

$$\alpha = 17.8\ 10^2\ d.\varepsilon'' \quad \text{dB/m for } 896 \text{ MHz} \qquad (2.5)$$
$$\alpha = 14.7\ 10^3\ d.\varepsilon'' \quad \text{dB/m for } 2450 \text{ MHz}$$

In general the loss factor of the material will vary with distance along the waveguide so that the fractional loss must be calculated for an interval Δz small enough to allow a constant averaged value of ε'' to be used. In common with r.f. applicators the power follows the variation of loss factor.

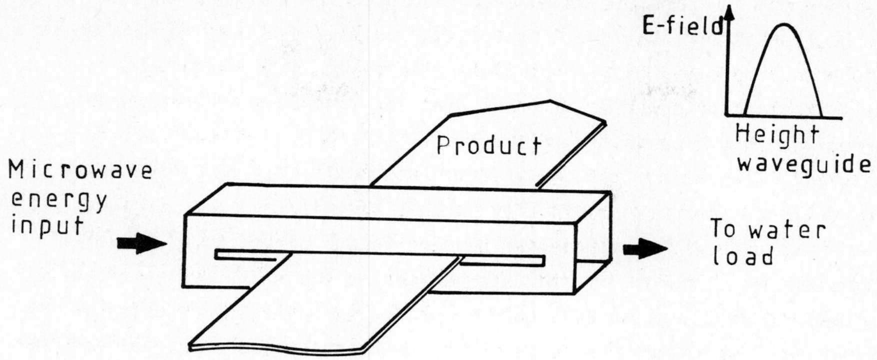

Fig. 7. Microwave travelling-wave applicator.

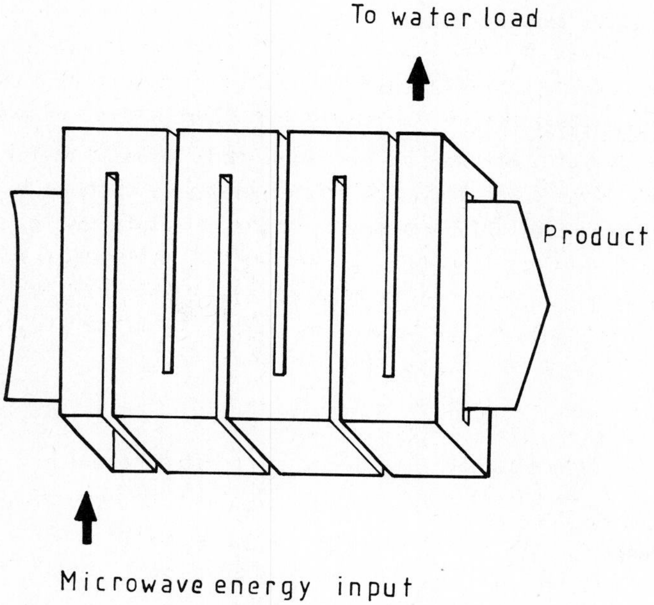

Fig. 8. Microwave travelling wave applicator in the form of a "serpentine".

To enable the equipment to be made more compact the waveguide can be doubled back on itself, to form a "serpentine" with the output terminated in a water load to absorb any energy not dissipated in the load, Figure 8. Typically the loss of power as given by equation 2.4 above in each "pass" of the serpentine should be in the region of 0.5 to 1.5 dB for good efficiency and uniformity of heating (ref. 3). If the attenuation is too large uneven heating results. The only possibility in this case is to use a straight length of waveguide with the material passing in the direction of travel of the electromagnetic energy. If the attenuation is too small resonant cavity methods must be used. As a rough guide, for drying applications using a travelling wave applicator the outlet moisture content should not be less than about 2% (ref. 9).

The dimensions of the waveguides mean that 2450 MHz is only suitable for sheet or web material whereas at 896/916 MHz aperture heights ~ 100mm allow material held on belts to be processed.

Single mode resonant cavity

A specialised form of applicator is the single mode resonant cavity which is capable of very large power densities over a relatively confined region and high operating efficiencies, Figure 9 (ref. 3). In principle the

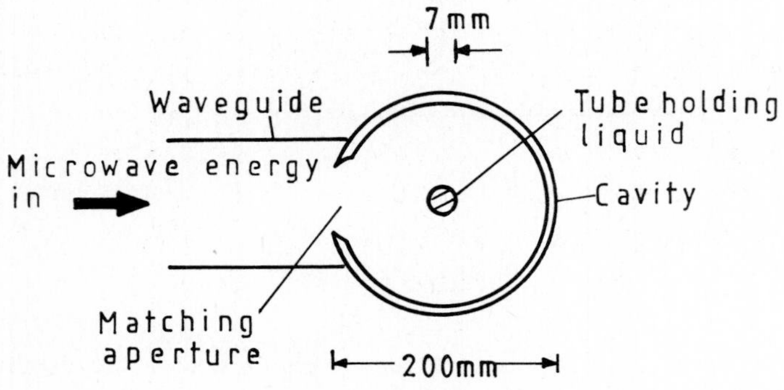

Fig. 9. Single mode resonant cavity operating at 896 MHz for heating liquids, plan view; power density ~ 10^8 W/m^3, E ~ 100 kV/m.

operation of these applicators is similar to the r.f. case in that they must be correctly coupled and tuned to the magnetron and coupling waveguide if power is to be dissipated into the load. Unlike the r.f. situation, when mismatching occurs power is still generated and must be dissipated in a dummy water load. Such applicators have been used for the rapid heating of liquids and for carpet drying (ref. 3).

2.5 The choice between r.f. or microwave equipment

The expression for the absorbed power per unit volume, equation (2.1), shows that for comparable values of loss factor and electric field strength much larger power densities ($\propto f$) can be achieved with microwave than with radio frequency methods. Alternatively, for the same power density the electric field at microwave frequencies is much smaller than that for radio frequencies and the possibility of electrical breakdown is smaller. The advantage of high power densities as the microwave frequency is increased is partially offset by the fact that the penetration depth decreases with frequency.

From previous sections it can now be seen that the choice of operating frequency in a given application is determined by:-

(a) the value of ε'' at the different frequencies,

(b) whether the process is batch or continuous,

(c) the size and geometry of the wet material.

When the values of ε'' permit the use of either radio frequency or microwave methods and the material dimensions are suitable for either, the determining factor will be whether or not the desired power density can be achieved at radio frequencies, bearing in mind that $E \lesssim 100kV/m$. If it can then the simpler mechanical construction and cheaper cost of radio frequency equipment will generally dictate the use of r.f. methods. In practice the two methods seldom have to compete against one another since some physical or operating consideration rules out one in favour of the other.

3. PHYSICAL BACKGROUND TO DRYING

The characteristics for drying with electromagnetic energy and hot air can be very different from those for hot air alone. This is due to the different physical mechanisms which determine the heat and mass flow. It is important to understand these differences because it is then possible to gain a qualitative understanding of the likely drying characteristics in a given situation even when the results from the mathematical model are rather limited.

The drying behaviour of a wet solid can depend on many factors, consequently a general, all embracing view of drying is impracticable. By considering the physical background to drying, a simplified classification of the drying regimes in which the use of dielectric heating is envisaged can be obtained in terms of a material's hygroscopic properties, its internal resistance to heat and mass flow, its overall dimensions and the principal source of energy available to provide the heat for evaporation. The drying characteristics obtained from theoretical modelling and experiments for materials within this classification provide a basis from which the drying behaviour of many wet solids can be assessed.

The purpose of this section is to set the scene for mathematical modelling and to highlight some important differences which can be expected between hot air and dielectric drying.

3.1 Hygroscopic Properties

Non-hygroscopic

If the pore size of the solid is greater than 0.1 μm and there is sufficient moisture so that the liquid molecules are not held in monolayers then the moisture will exert its full vapour pressure $P_{VSAT}(T)$ which is a function only of the temperature of the solid and is given by the Clausius Clapeyron equation - the material is non-hygroscopic. As such a solid dries out it will eventually show hygroscopic properties when the moisture is held on the pore walls as a film only a few molecules thick. This will arise when the solid is nearly bone dry.

Hygroscopic material

When the moisture and solid interact, for example, through hydrogen bonding, and/or the moisture is held in micropores of diameter less than 0.1 μm the liquid does not exert its full vapour pressure - the material is hygroscopic. It is customary to express the actual vapour pressure P_V as a fraction of the saturated value

$$P_V = \phi \; P_{VSAT}(T) \qquad (3.1)$$

where ϕ, the relative humidity, is a function of moisture content and temperature.

Further classification as a weak, moderate or strongly hygroscopic solid can be made according to the rate at which ϕ decreases with moisture content: the greater $\partial\phi/\partial X$ the stronger the moisture binding.

In addition to the latent heat of evaporation (r), further energy - the binding energy H_B - must be supplied to remove the bound moisture from the solid. Usually H_B is only significant for $\phi \lesssim 0.1$, with an upper value for very small ϕ of $H_B \sim r$.

3.2 Moisture distribution within the material

Pore radius $> 0.1\mu m$

When moisture is distributed continuously through the pores it is said to be in the "funicular state". Under these conditions liquid movement to the external surface of the material by capillary action can take place provided the pores are not too coarse. As moisture is removed the continuous threads of liquid gradually break up leaving isolated pockets of moisture and then capillary flow is only possible on a localised scale - the moisture is now in the "pendular state". When the material is close to bone dry (moisture content $\leqslant$ 1%) the molecules of moisture are held as a monolayer on the pore walls.

Pore radius $< 0.1\mu m$

As the pore size decreases the internal surface area can be very extensive and it is possible for the moisture to be retained on the pore walls as a film only a few molecules thick over the range of moisture contents typical for such hygroscopic materials.

3.3 Internal moisture movement

For hot air drying of a capillary porous body at atmospheric pressure the internal movement of moisture is commonly due to liquid flow to the surface by capillary action and vapour flow by molecular diffusion. The first flow can be related to the gradients in moisture content and temperature and the second to the gradient in partial vapour pressure or an equivalent temperature gradient. Less frequent forms of flow are surface diffusion of water molecules along the pore walls and at the reduced pressures used for freeze drying ($P \sim 100$ Pa) "Knudsen flow" where molecules move independently of one another.

In connection with dielectric drying two further mechanisms which occur as a result of the internal evaporation due to the volumetric energy absorption are important. These are filtrational flow of the liquid due to the excess internal pressure created in the gaseous phase, and convective flow of the gaseous phase due to an increase in the internal pressure of the air and vapour. Both these flows are proportional to the gradient in total pressure (air + vapour). Filtrational flow can only be expected when the material is saturated with moisture and its temperature is close to the boiling point, T_{BP}.

As a result of the non-uniform nature of the porous network, several of these flow mechanisms can be present at the same time. In particular liquid capillary flow in fine pores occurs alongside vapour diffusion in coarse pores. However, one will usually dominate during a given period of the drying process. This fact provides a useful simplification in the mathematical modelling. As the solid dries out and the moisture changes from the funicular to the pendular state and/or from non-hygroscopic to hygroscopic, so the flow will change from liquid to vapour movement.

3.4 Internal heat flow

Heat flow within the material occurs predominantly by conduction in the solid and liquid phases and is proportional to the temperature gradient. With dielectric drying there is a convective flow of heat in the vapour phase, this can usually be neglected in comparison with the other heat fluxes.

3.5 Internal resistance to mass flow

Provided the internal mass flow to the surface of the material can match the maximum rate of moisture removal which is possible from the surface to the airstream then the drying is close to that of a completely wet surface - this can apply even when the effective surface area for mass transfer is less than the actual surface area. The drying is controlled by the heat and mass balance at the surface of the solid. Although some information must be available to establish whether a particular material fits this category, detailed knowledge of the internal transport coefficients is not required. This situation is possible when the liquid is in the funicular state and moisture flow occurs by capillary action.

As the internal resistance to mass flow increases relative to the external resistance in the airstream so the drying rate becomes insensitive to the airstream conditions and is governed increasingly by the internal moisture movement.

When the moisture flow is due to vapour diffusion alone, the ratio of the internal resistance to flow relative to the resistance in the boundary layer is given by the Biot number Bi_M which in the case of a slab of thickness 2d is given by

$$Bi_M = \frac{k_M M_G d}{\rho_G D_V} \qquad (3.2)$$

where the value of the vapour diffusion coefficient for the porous body D_V is some fraction of that for free air. Typical values of $k_M M_G/\rho_G$ and D_V for a non-hygroscopic solid, are 10^{-2} - 10^{-1} m/s and 3.10^{-6} m^2/s respectively. If $Bi_M \leqslant 0.1$ the resistance to flow can be neglected, this requires that $d \leqslant 30$ μm. In general then, internal resistance to vapour flow by diffusion exists for all but the thinnest materials.

The convective flow of vapour due to the increase in internal pressure is by its nature independent of the external air stream and governed by the internal resistance to flow which for a slab of thickness 2d is given by

$$d/A_P \qquad (3.3)$$

where A_P is the pressure permeability. Typically, for pore radii $0.1 \leqslant R \leqslant 100$μm the range of values for A_P is $(10^{-6} \leqslant A_P \leqslant 10)$ m^2/s; the resistance to convective vapour flow is often negligible in comparison with that for vapour diffusion d/D_V.

Generally non-hygroscopic solids in which liquid flow by capillary action can occur will exhibit negligible resistance to mass flow down to relatively low moisture contents. Hygroscopic solids are more likely to exhibit internal resistance to moisture flow due to their smaller sized pores, the moisture-solid binding, and for some materials, particularly foodstuffs, their internal cellular structure and the presence of a shell.

3.6 Internal resistance to heat flow

For hot air drying, any internal resistance to heat transfer, due to the low thermal conductivity and/or the large overall dimensions of the wet solid, causes the surface temperature to approach that of the adjacent airstream so reducing the heat flow to the solid and creating significant temperature gradients.

The analogous Biot number for heat flow is given by

$$Bi_T = h_T d/k \qquad (3.4)$$

For $Bi_T \leqslant 0.1$ the temperature of the wet solid can be taken as uniform.

Typically, $10 \leq h_T \leq 100$ W/m^2°C, k~0.5 W/m°C so that for uniform temperatures $d \lesssim 10^{-2} - 10^{-3}$m. This condition is less restrictive than that for the mass flow by diffusion.

Not surprisingly, the internal resistance to heat flow has no direct effect on the volumetric absorption of electromagnetic energy. In contrast to hot air drying, temperature gradients are usually directed outwards to give heat flow to the surface where the energy is either used for evaporation or lost to the external airstream. For large values of Bi_T significant temperature gradients can exist. If, however, the heat losses are eliminated or reduced by ensuring that the airstream temperature is greater than that of the wet solid and/or by having a small value of h_T, the temperature gradients can be effectively eliminated. Under certain circumstances when electromagnetic energy is used to augment hot air drying the temperature gradients can be used to enhance liquid flow to the surface of the solid and thereby prolong the period during which the drying is controlled by the airstream and surface conditions rather than the internal resistance to mass flow.

3.7 Coupling of heat and mass transfer and the effects of the internal resistance to mass flow

3.7.1 Hot Air Drying

With hot air drying, heat and mass transfer are intimately linked. The slow movement of moisture towards the surface due to the internal resistance to mass flow results in the surface effectively drying out before the inside of the solid (for hygroscopic materials it will never be bone dry) and the solid increasing in temperature with the surface temperature approaching that of the adjacent airstream. The heat flow from the airstream to the surface of the solid drops and with it the rate of drying. As a consequence the 'dried out' surface of the material can reside at a relatively high temperature for much of the drying period while the moisture in the interior of the material is removed. This can lead to deterioration of the material due to chemical changes and shrinkage - "case hardening".

3.7.2 Drying with r.f. or microwaves

A feature of dielectric heating is that the heat and mass transfer are only partially linked: the increase in temperature due to the internal mass resistance does not affect the volumetric absorption of energy, it only influences the contribution from the airstream. Where the moisture removal is due to internal evaporation and vapour flow, the vapour is initially retained in the pores as a result of the internal resistance, causing the vapour pressure and temperature of the wet solid to rise. A situation is reached where the concentration of vapour is built up to a sufficient level for the gradients in the partial vapour pressure and total pressure (depending on the temperature reached) to produce a flow of vapour which can match the rate of internal vapour generation. In this sense the dielectric heating can always overcome the internal resistance to flow. This means that the possibility of case hardening can be reduced due to the shorter time that the surface is at a high temperature and the fact that with greater concentrations of vapour in the pores, as compared to hot air drying, in the case of hygroscopic solids the surface will be wetter. The quality of r.f. or microwave dried products is discussed later (see section 6.3).

The greater the resistance to vapour flow the higher the temperature and pressure at which the balance between evaporation and flow is achieved. This can lead to pressures at which the solid ruptures or swells and/or unacceptably high temperatures are reached. The first effect sets a limit on the power density which can be used while to overcome the second the solid must be cooled. This can be impracticable due to the large air flows required or the inefficient use of the absorbed electromagnetic energy. These features will be discussed more fully with the results obtained from the modelling.

4. APPLICATION OF DIELECTRIC HEATING TO DRYING: CLASSIFICATION OF DRYING REGIMES

From considerations of the dielectric properties of a wet solid and the heat and mass transfer it is apparent that r.f. and microwave heating are of potential benefit for removing unbound or weakly bound moisture in situations where conventional drying is unsatisfactory because of

a. inadequate heat transfer due to internal resistance to heat flow;
b. inadequate heat transfer due to small or non-uniform external heat transfer coefficients;
c. slow internal mass transfer when the moisture is held in the pendular state and/or the movement is by vapour flow;
d. the resulting poor quality of the dried product.

The removal of strongly bound moisture is not a common requirement. With hot air drying such materials exhibit characteristics similar to those for drying controlled by the internal resistance to mass flow. The application of r.f. or microwave techniques would depend on, the material having suitable dielectric properties, and the drying temperature which could be tolerated.

TABLE 4.1
Drying regimes considered in the text (denoted by X)

"Dimension" of wet material	Principal energy source	Hygroscopic Properties Non hygr./Weak	Hygroscopic Properties Strong	Internal resistance to heat flow	Internal resistance to vapour diffusion	Appropriate sections in text
Thin	Electromagnetic + hot air (depends on qb/h_T)	X	-	Negligible	Negligible	5.4.1, 5.6
Thick	Hot air (liquid pumping)	X	-	X	-	5.7.2
	Electromagnetic	X	-	X	X	5.4.2, 5.7.1
Particles held as a fixed bed	Electromagnetic (crossflow of air or forced flow absent)	X	X	X (in bed)	X (in bed or particle)	5.4.2, 5.8.1
	Electromagnetic + hot air (throughflow of air)	X	-	Negligible	Negligible	5.4.1,5.8.2 a,c,d,e
		-	X	Negligible	X (in particle)	5.4.2,5.8.2 b,c,d,e

This category of drying will be discussed in connection with porous particles.

It now remains to investigate the characteristics of drying with electromagnetic energy and hot air. In order to cover a wide range of drying operations three situations will be examined in which the dimensions of the wet material can be considered as (a) 'thin' $d < 0.1$mm (b) 'thick' $d > 10$mm and (c) a combination of these where 'thin' porous particles of diameter less than 10mm form a bed of thickness $2d \gtrsim 50$mm; these values of d are somewhat arbitrary. Cases (a) and (b) are intended to correspond to drying with negligible and appreciable internal resistance to heat and mass flow respectively. In these examples the moisture is taken to be only loosely bound with moisture flow generally occurring predominantly in the vapour phase. The accompanying air flow is across the external surface of the wet material.

With the drying of particulate matter the internal resistance to flow can be present in both the particles and the bed. The latter occurs when the air flow is across the surface of the bed. As will be seen, this configuration is essentially a variation of case (b). With hot air flowing through the bed any resistance to heat and mass flow associated with the actual drying (in contrast to the bed resistance to the air flow) is confined to the particles. Both non-hygroscopic and hygroscopic materials will be considered. Moisture flow can be in either phase at large moisture contents but is assumed to occur by vapour flow at lower values.

The operating pressure in each case above is atmospheric. In section 6.4 developments in vacuum and freeze drying will be discussed. The modelling for these situations can be derived from the equations applicable at atmospheric pressure, taking into account the different nature of the moisture flow which gives different values for the transport coefficients.

The classification of the drying regimes which will be considered is given in table 4.1. More details on the practical configurations will be given in the appropriate section.

5. MATHEMATICAL MODELLING

The purpose of mathematical modelling is to assist the practising engineer choose the most appropriate method of drying for a given product. To do this the theory must provide information on:

(a) the solid's drying temperature,

(b) the drying rate,

(c) the efficiency with which the input energy is utilised for sensible heating and evaporation,

(d) the state of the outlet air, that is, its humidity and temperature.

Other considerations such as product handling, plant size and environmental conditions are related to these factors. The detailed assessment of product quality must generally be obtained from analytical measurements on the dried product.

As seen from the previous section, several simplified drying regimes will be considered. After introducing the relevant transport equations some general guidelines on their simplification and solution are given. Significantly it is found that provided certain assumptions about the internal resistance to heat and mass transfer are valid, the broad features of the drying can often be anticipated on the basis of the relative proportions of the input energy supplied by the electromagnetic field and the hot air. The determination of the drying characteristics for thin and thick materials is then relatively straightforward. The drying of porous particles with internal resistance to flow is less amenable to such simplified modelling.

5.1 Modelling in perspective

At present the state of drying theory is such that it is unusual for generalised modelling to provide quantitative results for the drying of materials of industrial interest. Despite this, modelling is useful since it can give guidance, predicting qualitative features and defining specific regimes in which simplified equations can be used to provide quantitative information with a limited amount of data. Once confirmed by small-scale bench tests, these models can be used to scale-up results and examine the drying process over a range of conditions, a task which is often too time consuming by experimentation.

5.2 Heat and mass transfer equations for drying with electromagnetic energy and hot air

A wet porous solid is comprised of a number of discrete phases (solid, liquid and gas) distributed non-uniformly throughout the volume of the body. The formal derivation of continuity equations involves averaging quantities over small but finite volumes. The resulting equations are of little practical use until simplifying assumptions are invoked. For this reason most theories treat the porous solid as a continuous medium and use the familiar transport equations from continuum physics. This approach largely ignores the internal structure and distribution of the phases. The relatively simple mathematical appearance of the equations conceals the inherent complexity, due to the physical nature of the wet solid, within the transport coefficients which appear in the equations. These parameters cannot be derived from first principles, at best order of magnitude values can be estimated using simplified models.

5.2.1 Equations for internal flow

The standard continuum equations are due to Luikov (ref. 10) and can be written in the form:

Mass flow of moisture $$\gamma_o \cdot \frac{\partial X}{\partial t} = -\nabla j_M = -\nabla(j_V + j_L) \tag{5.1}$$

where $$j_M = -D\,(\nabla X + \delta_T\,\nabla T + \delta_P\,\nabla P)\,\gamma_o \tag{5.2}$$

Heat flow $$\gamma_o \frac{\partial}{\partial t}(Cps + XCp_L)T = \nabla k \nabla T + e_V\,\gamma_o\,r\,\frac{\partial X_L}{\partial t} + q \tag{5.3}$$

Total pressure $$\frac{\partial}{\partial t} C_a P = \nabla C_a\,A_P\,\nabla P - e_V\,\frac{\partial X_L}{\partial t} \tag{5.4}$$

These equations assume that the solid matrix does not change shape as the material dries out and that at any point the three phases are in equilibrium.

The moisture content X equals $X_L + X_V$ the mass concentrations of the liquid and vapour phase divided by the density of the dry skeleton solid, and e_V is the ratio of vapour flow to total moisture flow.

The presence of the electromagnetic energy absorption is seen in the heat flow equation and by the inclusion of the continuity equation for total pressure which is normally unimportant in conventional drying. The different mechanisms for moisture flow have been lumped together somewhat artificially into the coefficients D, δ_T, δ_P which are functions of moisture content and temperature and must be determined experimentally.

5.2.2 Boundary conditions at the surface of the wet solid

The general boundary conditions at the surface are:

Mass flow of moisture $$j_{MS} = M_L\,k_M\,\log_e\left[1 + \frac{P_{VS}-P_{VG}}{P_o - P_{VS}}\right] \tag{5.5}$$

The right hand expression, derived from film theory (ref. 11), includes the diffusive and convective flow of vapour across the external body layer. Removal of liquid moisture by mechanical effects is not included.

As $P_{VS} \rightarrow P_o$ the expression above becomes indeterminate and the moisture flow which is entirely convective is obtained from the total pressure equation.

Heat flow $h_T \, \beta_T(T_G - T_S) - k\nabla T - rj_L = 0$ (5.6)

with ∇T and j_L evaluated at the surface.

The first term is the amount of heat brought to the surface of the body from the air stream, the second the amount of heat passed into the body and the third term the amount of heat expended in evaporating the liquid flow at the surface. The correction factor β_T accounts for the decrease in heat transfer which occurs when the boundary layer is destablished by the convective flow of the evaporating moisture (ref. 12) and is given approximately by

$$\beta_T \sim 1 - \frac{P_{VS}}{P_o} \qquad (5.7)$$

Total pressure $P_S = P_o$ (5.8)

The total pressure at the surface is equated to atmospheric pressure as a result of the fact that any excess pressure is relaxed at the velocity of sound.

5.2.3 Heat and mass transfer equations for the air stream

The evaporation of moisture into the airstream and the flow of heat by conduction across the external boundary layer mean that the humidity and temperature of the air change through the dryer. The air flow can be idealised as either plug or perfectly mixed flow. The expressions for the change in humidity and temperature for an incremental distance dz through the dryer for plug flow are:

$$dY_G = \left[\frac{qb}{h_T} + \beta_T(T_G - T_S)\right] \frac{Cp_G}{r} N_T \frac{dz}{Z_o} \qquad (5.9)$$

$$dT_G = -\beta_T(T_G - T_S) \, N_T \frac{dz}{Z_o} \qquad (5.10)$$

where b is the ratio of the volume for electromagnetic energy absorption to the external area for heat and mass transfer and N_T is the number of transfer units, $h_T a Z_0/GCp_G$.

For perfectly mixed air flow Y_G and T_G undergo a stepwise change after entering the dryer and thereafter their values remain constant at Y_{GOUT}, T_{GOUT}.

These are given by

$$Y_{GOUT} = Y_{GIN} + \int \frac{j_{MS} a\, dz}{G} \tag{5.11}$$

$$T_{GOUT} = T_{GIN} - \int \frac{\beta_T h_T\, (T_{GOUT} - T_S)\, a\, dz}{GCp_G} \tag{5.12}$$

When quasi-steady conditions are assumed throughout the dryer for the temperature of the wet solid, that is, the sensible heat term can be ignored, the overall energy balance on the dryer can be expressed in the form

$$\frac{Q}{GCp_{GIN}} + (T_{GIN} - T_{GOUT}) = \frac{r_{OUT}}{Cp_{GIN}} (Y_{GOUT} - Y_{GIN}) \tag{5.13}$$

$$\text{with} \quad Y_{GOUT} = \frac{0.622\, \phi_{GOUT}\, P_{VSAT}\, (T_{GOUT})}{P_o - \phi_{GOUT}\, P_{VSAT}\, (T_{GOUT})} \tag{5.14}$$

while the power available for evaporation is given by

$$Q + G(Cp_{GIN}\, T_{GIN} - Cp_{GOUT}\, T_{GOUT}) \tag{5.15}$$

assuming no heat losses.

In connection with the drying of particles with electromagnetic energy and a throughflow of air, the equations for heat and mass transfer at the surface of a particle for quasi-steady conditions can be combined to give

$$\frac{qb}{h_T} + \beta_T\, (T_G - T_S) = M_L r\, \frac{k_M}{h_T} \log_e \left[1 + \frac{P_{VS} - P_{VG}}{P_o - P_{VS}} \right] \tag{5.16}$$

with b the volume to area ratio for an individual particle.

5.3 Simplification of the drying equations using physical considerations

The solution of the heat and mass transfer equations in the form above taking into account the variation of the coefficients with moisture content and temperature, is generally both impracticable and impractical due to the lack of physical data and the considerable effort required to solve the equations numerically (ref. 13). Only in specific cases where useful approximations cannot be made or where the nature of the drying operation requires accurate predictions of the drying parameters, as for example in freeze drying, is such an approach required.

In practice it is necessary to look for some simplifying assumptions which make the equations more manageable and reduce the amount of data required. The main problem encountered is how to deal with the internal resistance to mass transfer and its effect on the moisture distribution within the wet solid.

The physical nature of drying with electromagnetic energy, as discussed in section 3, and the operating regimes of interest suggest the following simplifying assumptions:

a. negligible internal resistance to heat and mass flow, $D = k \to \infty$;

or

b. infinite resistance to vapour flow by diffusion, $D_V \to 0$.

Implicit in each of these assumptions is the quasi-steady condition that $\partial T_S/\partial t = \partial X_V/\partial t \approx 0$ at some stage during the drying.

The conditions under which (b) can be used and the characteristics which result from these simplifications are related to the ratio of the electromagnetic energy absorbed by the wet solid to the heat transferred to or from the airstream. By first examining the drying in a general way as a function of this ratio, the characteristics for the three cases of practical interest can be readily appreciated.

5.4 General drying characteristics

5.4.1 Negligible internal resistance to heat and mass flow

The characteristics obtained in this case can be found by considering the drying of a volume V of a non-hygroscopic solid with external surface area A. The power absorption q and airstream conditions are taken as constant - the latter would apply for ideal mixing of the air. The conclusions obtained apply equally well for hygroscopic materials, plug flow of the air and q varying during the drying.

The form of the internal mass flow does not make a significant difference to the final transport equation obtained and so for convenience $e_V = 0$, i.e. liquid flow only, is assumed. The heat and mass transfer equations reduce to

$$\frac{b\gamma_0\,(Cps + X_L\,Cp_L)}{h_T}\,\frac{\partial T_S}{\partial t} = \frac{qb}{h_T} + \beta_T(T_G - T_S) - M_L\frac{rk_M}{h_T}\,\log_e\left[1 + \frac{P_{VS} - P_{VG}}{P_o - P_{VS}}\right] \qquad (5.17)$$

To solve this equation the dielectric and hygroscopic properties must be known and iterative techniques would be required to establish the values of T_G and P_{VG} or equivalently the humidity Y_G. The method of solution is straight forward and it is not essential to invoke condition (a), however its use does allow a quick estimate of the solids' temperature and the overall drying time to be made. Equation 5.17 has been solved for assumed values of Y_G and T_G with q constant during the drying, using qb/h_T as a variable. The significance of this parameter lies in the fact that

$$\frac{qb}{h_T\beta_T(T_G-T_S)}$$

is the ratio of the heat transfer due to electromagnetic energy absorption compared to that due to conduction from the airstream. The values of T_S at which $\partial T_S/\partial t = 0$ are shown in Table 5.1. Several regimes are apparent according to the value of qb/h_T.

TABLE 5.1
Variation of surface temperature T_S with qb/h_T °C, $\beta_T = 1$.

	T_G = 40°C		T_G = 120°C	
qb/h_T °C	Y_G = 0.005 kg/kg	Y_G = 0.05 kg/kg	Y_G = 0.005 kg/kg	Y_G = 0.05 kg/kg
	T_S °C		T_S °C	
0	19.6	40.0	37.5	49.0
10	22.7	41.9	39.0	50.0
20	25.5	43.0	40.4	50.8
50	32.3	46.0	44.0	52.8
100	40.4	50.8	49.0	56.0
200	50.7	57.3	56.0	61.2
500	65.6	68.5	67.9	70.4
1000	76.0	77.3	76.5	79.2
2000	84.3	84.8	84.7	85.2
5000	91.4	91.6	91.5	91.64

At large values $qb/h_T \gtrsim 500$°C the drying characteristics are due principally to the electromagnetic energy alone. The drying is insensitive to T_G, Y_G and to a lesser extent h_T, with the heat transfer to or from the airstream equal to or less than 10% of the energy used in evaporation. The primary role of the airstream is to remove the evaporating moisture from the dryer. As qb/h_T increases the temperature T_S rises towards the boiling point temperature T_{BP}. The time taken to reach the equilibrium value of T_S can be a significant part of the total drying time. The rate of evaporation, however, only becomes significant as T_S approaches its equilibrium value.

Once the value of T_S for $\partial T_S/\partial t = 0$ is known the overall process time can be estimated from the total energy necessary for the initial sensible heating and moisture evaporation and the total electromagnetic energy supplied to the dryer, using the assumption that the contribution from the hot air is negligible.

At small values $qb/h_T \leqslant 50°C$ the drying is sensitive to changes in humidity and temperature Y_G and T_G. The hot air contributes to the evaporative energy. The temperature of the solid T_S is within about 10°C of that for hot air drying alone. Following an initial fast increase in T_S the drying rate quickly reaches an approximately constant value while $\partial T_S/\partial t$ becomes small. For practical purposes $\frac{\partial T_S}{\partial t} \approx 0$ can be assumed from the start of the drying.

For intermediate values $50°C < qb/h_T < 500°C$ the behaviour changes from the hot air to the electromagnetic dominated drying characteristics as qb/h_T increases. Equation (5.17) should be solved with $\partial T_S/\partial t$ retained.

5.4.2 Significant internal resistance to vapour diffusion

In this case the vapour concentration builds up until the internal mass flow can match the rate of internal evaporation and $\partial X_V/\partial t = \partial T/\partial t \approx 0$. When this condition applies equations (5.1)-(5.3) give after some rearrangement and substitution,

$$\gamma_0 \Delta X_V \sim \left[\frac{qb}{h_T} - \beta_T\,(T_S - T_G)\right]\left(\frac{Cp_G}{r}\right) \rho_G\, Bi_M \tag{5.18}$$

where $\gamma_0 \Delta X_V$ is the difference in vapour concentration at the centre and surface of the wet solid and $Bi_T < 0.1$ is assumed for simplicity. The variations in the drying behaviour as the value of $\frac{qb}{h_T}$ changes can be obtained from this equation.

Neglecting the heat loss term $\beta_T(T_G - T_S)$ for the moment, it can be seen that depending on the value of Bi_M, equation (5.18) implies large values of vapour concentration at the centre of the material to give values of T_S around or in excess of 100°C. For example with $Cp_G/r \sim 1/2500$, $qb/h_T =$ 10°C, $Bi_M = 100$ then $\gamma_0\ \Delta X_V$ is equal to 0.4 kg/m^3 and $T_S \sim 90°C$.

At such temperatures the convective vapour flow mechanism would come into operation. Provided the resistance to this flow b/A_P is small the pressure within the pores does not build up significantly, $T_S \sim T_{BP}$, the boiling point of the liquid, and the quasi-steady approximation $\partial T_S/\partial t = \partial P/\partial t \approx 0$ is applicable.

From an initial situation with internal resistance to mass flow the electromagnetic heating is able to create a condition of negligible resistance to flow. This is achieved at the expense of the solid's temperature which is raised well above that for hot air drying.

Progressing from this simple picture it is now necessary to take into account the influence which the heat losses have. For this purpose assume that T_S = 100°C is reached, that T_G = 50°C and $\beta_T \sim 1$ for simplicity. The three ranges of qb/h_T will be considered as before.

For $qb/h_T \geqslant$ 500°C heat losses are insignificant and the air stream has little influence on the heat transfer. With $Bi_M \gtrsim 10$ the temperature T_S reaches 100°C and the convective flow dominates. The picture presented above is essentially unaltered. In this case it can be assumed that $D_V = 0$ and $\partial X_V/\partial t = \partial X_L/\partial t \approx 0$ until $T_S = T_{BP}$ is reached.

The drying can be simplified into three stages: initial sensible heating, a period during which the maximum pressure is attained, which is negligible for small b/A_P, and the final moisture removal period (ref. 14).

Turning to the lower range $qb/h_T \lesssim$ 50°C the postulated conditions T_S = 100°C for T_G = 50°C cannot apply since they would make the heat losses greater than the absorbed electromagnetic energy: this result indicates that for these values of qb/h_T the airstream actively participates in the heat transfer. The surface temperature T_S is sensitive to the particular combination of T_G and Bi_M. For $T_G + qb/h_T <$ 100°C an upper limit can be placed on the temperature T_S: T_S tends to $qb/h_T + T_G$ as Bi_M increases. In these circumstances much of the applied electromagnetic energy can be dissipated as heat losses to the airstream. When T_G or T_G + qb/h_T is greater than 100°C the value of T_S approaches 100°C as Bi_M increases, and with hygroscopic solids temperatures in excess of this can be obtained. These points are illustrated in table 5.2 which shows calculated values of T_S under various conditions for a non-hygroscopic solid.

To model the situation $qb/h_T <$ 50°C the quasi-steady condition can be assumed from the beginning as far as moisture removal is concerned, and for $Bi_M \gtrsim 10$ the surface vapour pressure can be taken as equal to that of the adjacent airstream with the mass flow given by $-\gamma_0 D_V \nabla X_V$. This enables a value of $\gamma_0 X_V$ at the surface of the material, and hence T_S, to be obtained from equation (5.18) provided Bi_M is known.

TABLE 5.2
Drying wich internal resistance to vapour diffusion, effect of T_G and Bi_M on the temperature T_S for low and intermediate values of qb/h_T. $\gamma_o X_V$ in the airstream = 0.006 kg/m^3

T_G °C	qb/h_T °C	Bi_M		
		0	10	100
		T_S°C		
	10	26	45	57
50	50	34	62	83
	100	42	73	92
	10	36	65	86
100	50	52	73	100
	100	47	100	100

In general, however, data on the internal mass transfer are lacking and a qualitative approach must be adopted. The internal resistance for non-hygroscopic materials can be represented as a decrease in the effective surface area for mass transfer. The drying of hygroscopic materials with and without internal resistance to mass flow can be modelled using the surface conditions, that is the relative humidity, as a variable. Temporal changes cannot be obtained explicitly.

In the intermediate range 50°C $<$ qb/h_T $<$ 500°C with $Bi_M > 10$ the drying characteristics rapidly tend towards the electromagnetic energy dominated regime with the air playing a secondary role. In general, equation (5.18) can be used to estimate T_S and depending on its value the appropriate limiting condition corresponding to large or small values of qb/h_T can be taken.

5.5 General approach to the solution of the drying equations

For a given drying process, the first quantities to be determined are q, b and h_T. Order of magnitude values of h_T can be found from correlations while a value of q can be estimated from the total electromagnetic power available as determined from the energy required, assuming that the hot air contributes nothing, or alternatively, from some preset value of the amount of r.f. or microwave energy to be used.

Once a rough idea of the value of qb/h_T is known, and an appropriate choice of simplifying assumptions for the internal resistances to heat and mass flow is made, it is then possible to form a view of the likely drying characteristics.

5.6 The drying characteristics of thin sheets and webs

The conventional drying of thin sheets and webs often produces irregular moisture profiles across the width of the sheet due to uneven external heat transfer coefficients. The selective energy absorption properties of electromagnetic techniques can be exploited to reduce the differences in moisture profile due to the initial conventional drying. In addition, the rate of drying of the loosely bound moisture held in the pendular state can be significantly increased. The type of applicator used for such a continuous process would be an r.f. sytem with staggered throughfield or strayfield electrodes or less commonly a microwave serpentine waveguide. Electric field strengths $\sim$ 100 kV/m would be possible.

Typical physical parameters are $k \sim 0.1$ W/m°C, $D_V \sim 10^{-6}$ m^2/s, $h_T \sim 10$ W/m^2°C, $k_M\ M_G/\rho_G \sim 0.01$ m/s, $\varepsilon'' = 1$, so that for a thickness 2d = 0.2mm, $Bi_T < 0.1$, $Bi_M \sim 1$ and over the frequency range 13.5-2450 MHz, 60°C $< qb/h_T <$ 10^4°C. Although the mass Biot number is greater than 0.1, the internal resistance to mass flow is not large and in the first instance can be neglected. The range of values of qb/h_T covers both the hot air and electromagnetic dominated drying regimes. Consequently in order to predict the drying behaviour some idea of q in each particular case must be known. A specific example of moisture levelling of paper webs will be considered.

The relevant heat and mass transfer equation for vapour flow $e_v = 1$ is given by equation (5.17) with $b\gamma_o r\ \partial X_V/\partial t$ added to the left-handside (this term is not in fact significant in comparison with the sensible heat term). The solution of this equation using measured dielectric and hygroscopic data obtained at elevated temperatures, has been carried out for r.f. paper drying (refs. 15 and 16). A typical result for two webs initially at different moisture contents is shown in Figure 10a, while in Figure 10b an experimentally measured curve is shown for comparison. The peak values of q used in this work gave values of $qb/h_T \sim 10^3$°C. The moisture levelling effect is clearly seen. The close agreement between experimental and computed drying curves demonstrated that the assumption of negligible internal resistance to mass flow was justified.

The characteristic of moisture levelling is that $\partial X_L/\partial t$ should be proportional to X_L: the moisture content changes fastest at the wettest

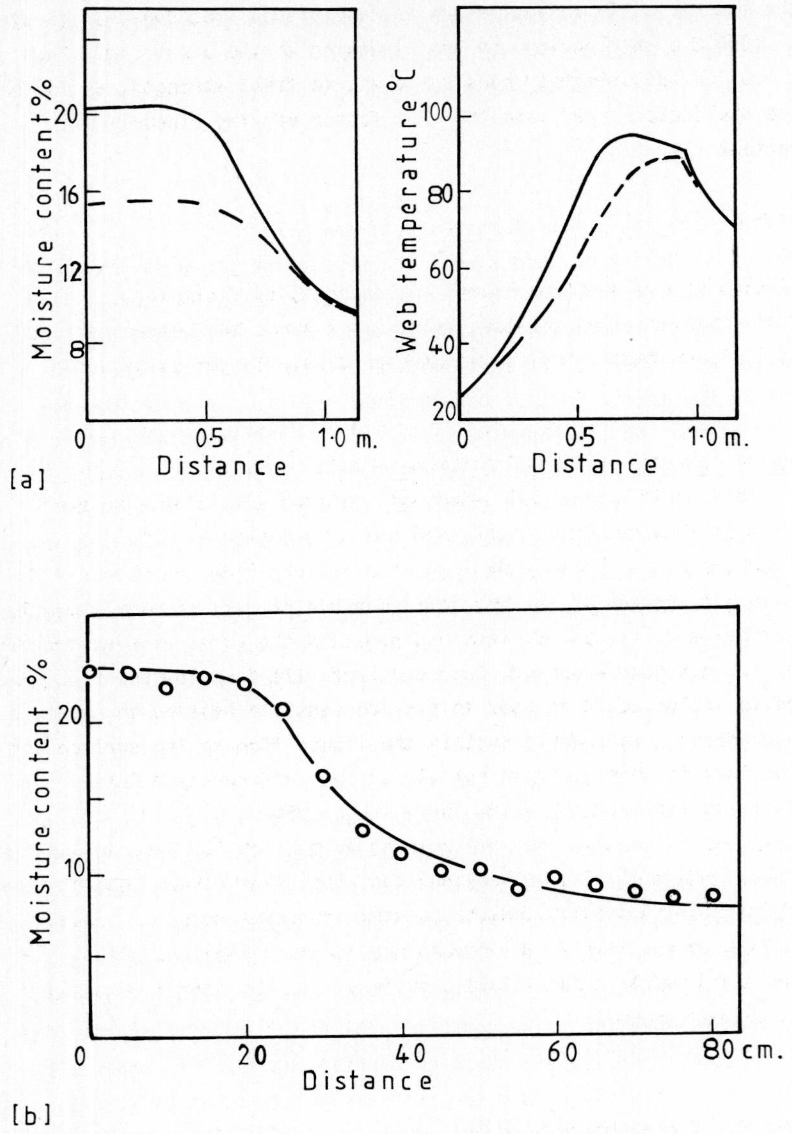

Fig. 10. Drying of paper webs at 27.12 MHz, (a) typical theoretical results for two webs initially at different moisture contents, (b) a typical experimental result. Distance is measured along the direction of travel of the webs from the initial point of application of the r.f. heating (after ref. (15) with permission of Dr. P. L. Jones).

parts of the material. This feature can be easily seen from equation (5.3) for $qb/h_T \sim 1000°C$ by neglecting the heat transfer to the airstream. Then for $\partial T_S/\partial t \approx 0$, an approximately constant electric field strength, as is the case for the applicators used, and the loss factor varying linearly with moisture content

$$\gamma_o \frac{\partial X_L}{\partial t} = \frac{-q}{r} \propto \varepsilon'' \propto X_L \qquad (5.19)$$

The variation of q with temperature, neglected in this simple derivation, further enhances the levelling effect since the wetter parts of the web reach higher temperatures and correspondingly larger values of q.

5.7 The drying characteristics of thick materials

With materials of thickness $2d > 20mm$ both internal resistance to heat transfer and mass flow by vapour diffusion can be expected $Bi_T > 1$, $Bi_M > 30$ (see sections 3.5 and 3.6). As long as capillary flow can maintain liquid flow to the surface of the wet solid the use of r.f. or microwaves to provide evaporative energy is not required provided the external heat transfer is adequate. It has been proposed, however, (ref. 17) that low power electromagnetic fields could be used to provide sensible heating to create a temperature gradient; this would sustain the liquid flow to the surface for a longer period than is possible with hot air drying and thereby maintain the drying rate at its initial high value for a longer time.

Where the pores of the material are coarse, as applies for example in the drying of most "packages" of textile yarn, capillary flow cannot be sustained even at large moisture contents and the moisture movement occurs by vapour flow. In this instance, r.f. or microwave drying using parallel plate electrodes or a multimode cavity would provide all of the energy required. This case is treated first.

5.7.1 Drying at large power densities

For $E_{max} \sim 100$ kV/m, $\varepsilon'' \sim 1$, $h_T \sim 10$ W/m^2°C and $d > 10$mm the range of maximum values of qb/h_T is 750-1.36 10^6 °C. The approximation of infinite internal resistance to vapour diffusion leading to $\partial X_L/\partial t \simeq 0$ until the temperature reaches T_{BP} can be used. The time to reach T_{BP} is given by

$$t_{BP} = \frac{\gamma_o(Cps + XCp_L)\,(T_{BP} - T_{INITIAL})}{q_{BP}\,(1-g)} \log_e \frac{1}{g} \qquad (5.20)$$

where $g = \frac{q_{initial}}{q_{BP}}$ is the ratio of the initial and final values of q respectively. The value of g is about five at microwave frequencies while for radio frequencies $g < 1$.

For mass flow to two surfaces the maximum pressure generated at the centre of the material is obtained from equation (5.4) with $\partial P/\partial t = 0$:

$$P_{max} - P_o = \frac{1}{2\, C_a\, \gamma_o} \frac{q_{BP} d}{r} \frac{d}{A_p} \tag{5.21}$$

where $C_a \sim \frac{M_L\, \Phi}{\gamma_o R_G (T + 273)}$

and an order of magnitude value of A_p is found from

$$A_p \sim \frac{P_o\, R^2}{8\mu} \tag{5.22}$$

for steam at 100°C, with viscosity μ, held in parallel pores of radius R. This pressure permeability A_p increases with decreasing moisture content. The extension of equation (5.21) for drying in three dimensions is straight forward.

For fine pored materials of diameter $< 10\mu m$ and large thickness the value of d/A_p may impose restrictions on the permissible power density if some critical pressure is not to be exceeded. This happens, for example, with the drying of wood and ceramics where cracking and distortion can occur if too much power is applied (ref. 18).

On the assumption of negligible internal resistance to the convective vapour flow the rate of moisture removal is given by

$$j_{MS} = -\gamma_o\, C_a\, A_p \left.\frac{\partial P}{\partial z}\right|_{z = \pm d.} \tag{5.23}$$

which leads to

$$\gamma_o \frac{\partial X_L}{\partial t} = - \frac{q_{BP}}{r} \tag{5.24}$$

Once again if the absorbed power is proportional to the moisture content (r.f. systems) levelling of initial uneven moisture profiles takes place.

With materials saturated with moisture, convective liquid flow can be obtained as the temperature approaches the boiling point T_{BP}. The duration of this flow depends on the way the moisture redistributes itself as liquid is removed and can really only be found by experiment. The occurrence of this "liquid pumping" is very dependent on the product and its moisture content.

In particular instances it has been used to great effect to reduce the overall energy requirements for moisture removal (ref. 19).

Experimental results for the microwave drying of cotton yarn wound on a central spindle to give a cylindrical package 200mm in diameter are shown in Figure 11. The initial decrease in moisture content for curve 1 is due to liquid pumping. Under other conditions where the initial moisture content was smaller, curve 2, this phase was not seen and little moisture removal occurred until the temperature neared the boiling point of water. Once T_{BP} is reached the drying follows the expected trend with the moisture removal proportional to the input power (ref. 20).

An important effect was observed in related experiments (ref. 21) concerned with the drying of cotton fabric containing "finishes" dissolved in the liquid: there was a lack of migration of these finishes during the microwave drying. This was in contrast to the hot air drying where the initial capillary flow to the surface resulted in an uneven distribution of the deposited finish.

5.7.2 Drying at low power densities

The principle behind the proposed use of small power densities ($qb/h_T < 10°C$) to increase the liquid mass flow by capillary action to the surface of the solid can be seen from equations (5.1) to (5.6) assuming quasi-steady conditions and the absence of internal evaporation $e_V = 0$. The effect relies on the additional flow $-D_L\gamma_o\delta_T\nabla T$ producing a zero gradient in liquid concentration at the surface, $\nabla X_L = 0$. For one dimensional flow the heat and mass balance at the surface gives

$$-D_L\gamma_o \frac{\partial X_L}{\partial z} = \frac{h_T}{r} \beta_T (T_G - T_S) - \frac{\partial T}{\partial z}\left(\frac{k}{r} - \gamma_o\delta_T D_L\right) \tag{5.25}$$

where $\partial T/\partial z$ is negative. Since the effect of the electromagnetic energy is assumed to be small in comparison with the hot air, the value of the first term on the right is taken to be that for the hot air drying alone. It is seen that a necessary but not sufficient condition for $\partial X_L/\partial z = 0$ is that $k/r < D_L \gamma_o\delta_T$: the increase in the rate of evaporation at the surface of the solid due to the heat conducted from the centre must be less than the increase in mass flux to the surface. Without this condition the liquid pumping at low power densities will not be seen.

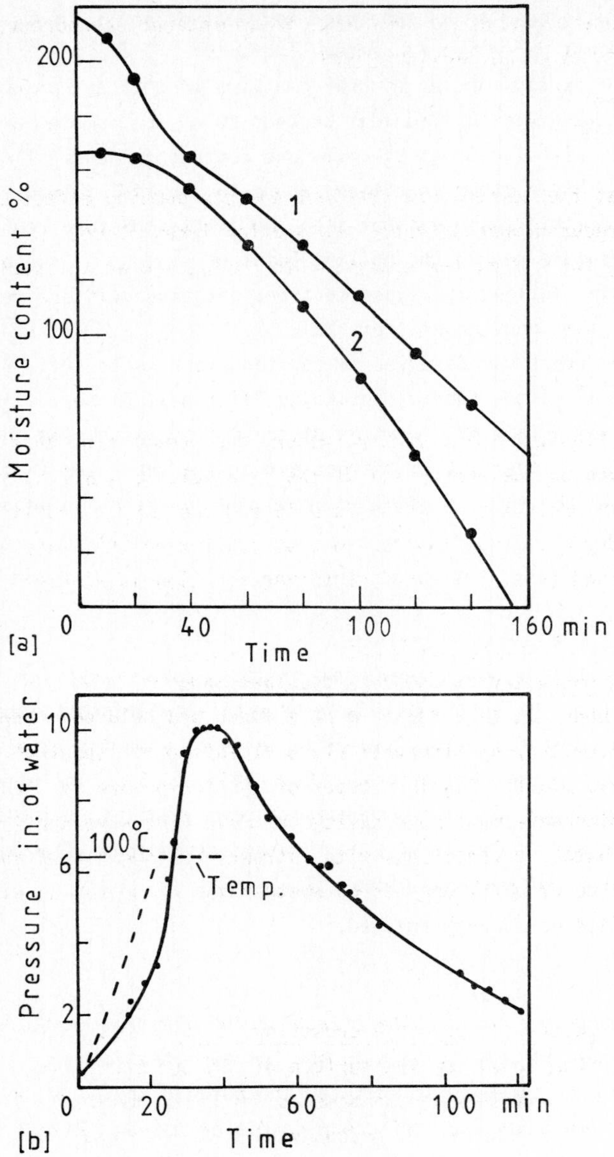

Fig. 11. Experimental results for the drying of cotton yarn with microwaves, (a) moisture content versus time; (b) pressure and temperature near the centre of the package versus time, corresponding to the drying conditions for curve 1 (after ref. (20) with permission of Pergamon Press).

A further consideration is the change which internal evaporation causes $e_V \neq 0$. The steady-state balance gives

$$-k\nabla^2 T = q + e_V \frac{\partial X_L}{\partial t} \tag{5.26}$$

which shows that the temperature gradient and the pumping effect due to this mechanism are reduced when internal evaporation takes place.

Substituting for $\partial T/\partial z$, the condition for $\partial X_L/\partial z = 0$ at the surface, with $e_V = 0$, is

$$\frac{qd}{h_T} = \frac{\beta_T(T_G - T_S)}{(D_L\gamma_o\delta_T r/k - 1)} \tag{5.27}$$

As with the liquid pumping at high power, the presence of this effect depends very much on the properties of the material, D_L, δ_T, k, the initial state in which the moisture is held and the way it redistributes itself during drying. Experiments aimed at confirming this pumping action have been reported (ref. 17) but further verification is required.

5.8 The drying characteristics of particulate beds

Particulate beds can take the form of a fixed bed held on a band, perforated belt or tray or alternatively a fluidised or 'spouted' bed in which the particles are mixed. Possible types of applicator are r.f. parallel plate electrodes, a microwave multimode cavity or travelling wave applicator. The process can be batch or continuous with microwaves being preferred for batch drying. The value of qb/h_T depends on whether the accompanying air flow is across the surface or through the bed.

5.8.1 Electromagnetic energy and a crossflow of air

When the air flow is across the surface of the bed the drying characteristics are essentially the same as for thick materials at large values of qb/h_T. For powders and granules with $h_T \sim 20$ W/m^2 °C, $2d \gtrsim 50$mm, $\varepsilon'' \sim 0.5$ the range of the maximum values for qb/h_T using $E_{MAX} \sim$ 100kV/m is $10^3 - 10^7$ °C. In practice electric field strengths below E_{MAX} are used so that actual values of qb/h_T can be an order of magnitude smaller. Provided the diameter of the particles is $\gtrsim 100\mu$m the bed itself offers little resistance to the convective flow of vapour and negligible pressure is built-up in the bed.

Since the particles do not form a rigid bed any pressure built up in beds of very fine particles results in 'channelling' throughout the bed which provides paths which allow the pressure to be relieved. If the individual particles have internal resistance to mass flow then rupturing can occur, since their temperature will be at or above the boiling point T_{BP}.

Extending the earlier analysis to cover hygroscopic materials, the moisture removed is again assumed to be negligible until $T = T_{BP}$ after which the quasi-steady approximation $\partial T/\partial t \approx 0$ is used. The rate of total moisture removal from the bed is given by

$$\gamma_o \int \frac{\partial X_L}{\partial t} \, dV = - \int k_M \, M_L \, \log_e \left[1 + \frac{P_{VS} - P_{VG}}{P_o - P_{VS}} \right] dA \approx \int \frac{-q \, dV}{r + H_B} \tag{5.28}$$

As the solid dries out and P_{VS} decreases with moisture content, the temperature rises in an attempt to compensate and maintain the partial vapour pressure at a value consistent with equation (5.28).

Drying and temperature curves obtained by the author from microwave experiments with a fixed bed of depth $\sim$ 50mm are shown in Figure 12 (curves marked 1) for alumina beads, a non-hygroscopic material with negligible internal resistance to mass flow, and silica gel, a strongly hygroscopic material with internal resistance to mass flow. The observations are as predicted.

5.8.2 Electromagnetic energy and a through-flow of air

If air is passed through the bed, caking of the particles can be eliminated and there is the possibility of drying at lower temperatures as compared with the cross-flow case. By using a fixed bed rather than a fluidised bed attrition of particles can be prevented. This case is considered below for batch drying. Situations where fluidising techniques are applicable will be discussed in section 6.4.1.

The relevant dimension for determining the parameter b is the diameter of the particle; $h_T \sim 100$ W/m^2 °C and the electric field strengths used are certainly below E_{MAX}. Typical values of qb/h_T are $\lesssim 20$°C: the airstream can be expected to participate in the heat transfer. The quasi-steady approximation $\partial T_S/\partial t \approx 0$ is used and plug flow is assumed for the air stream. Modelling of this situation requires the simultaneous solution of the transport equations for the particles and the air stream, equations (5.11), (5.12) and (5.16).

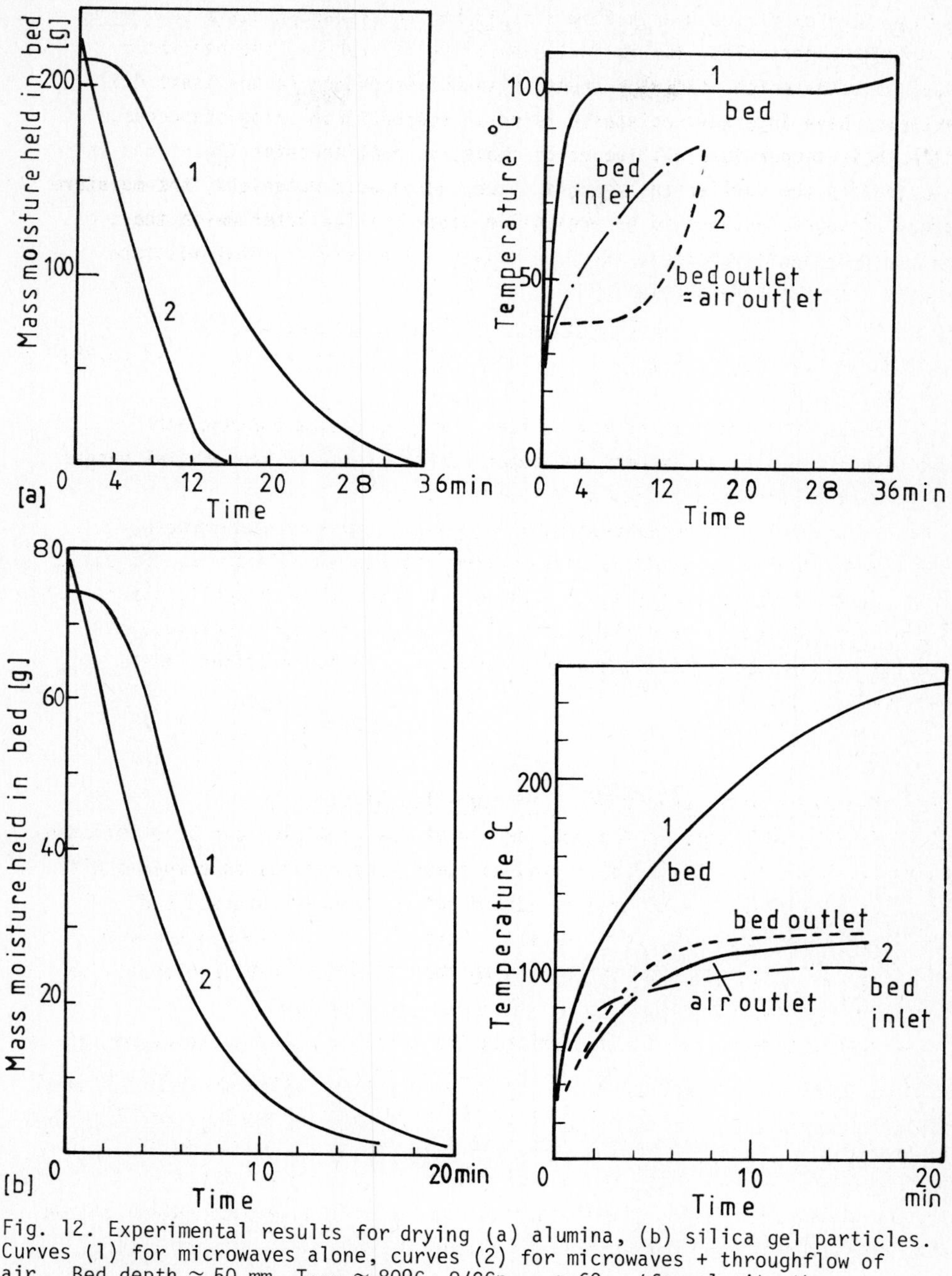

Fig. 12. Experimental results for drying (a) alumina, (b) silica gel particles. Curves (1) for microwaves alone, curves (2) for microwaves + throughflow of air. Bed depth ~ 50 mm, $T_{GIN} \simeq 80^{o}C$, $Q/GCp_{GIN} \sim 60$. (for clarity the experimental points are not included)

(a) Non-hygroscopic solid, negligible internal resistance to heat and mass flow

The variation of the air stream humidity and temperature, and solids' temperature through a fixed bed is shown in Figure 13 for a constant value of q throughout the bed. The distance is measured from the dryer inlet and is normalised by the bed depth. Such a situation would be obtained in the early stages of batch drying in a microwave oven (q = constant) before any part of the bed dries out (ref. 22).

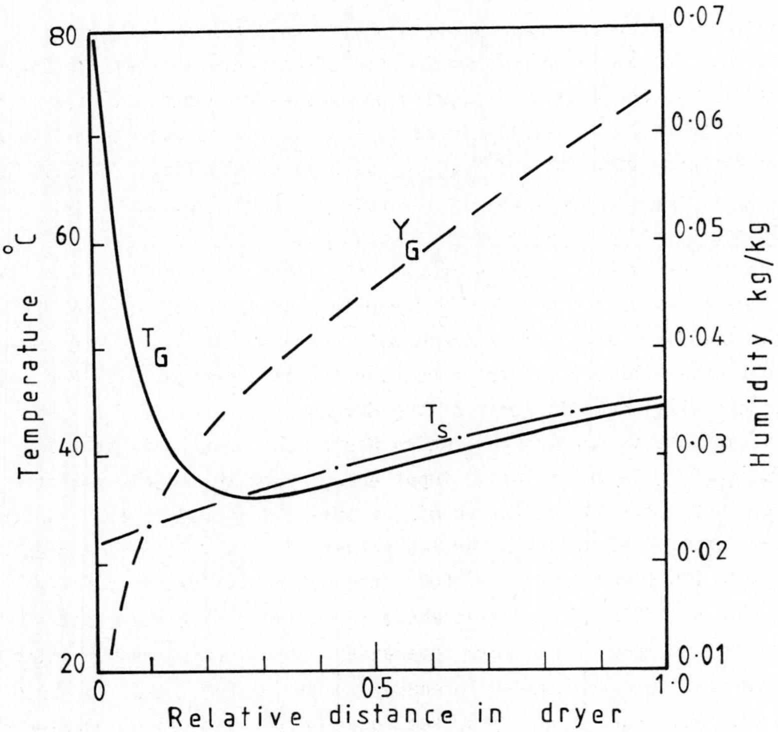

Fig. 13. Computed variations in airstream humidity Y_G and temperature T_G, and solids' temperature T_S as a function of relative distance through the fixed bed ref. (22) T_{GIN} = 80°C, Y_{GIN} = 0.005 kg/kg, $\frac{qb}{h_T}$ = 10, N_T = 10.

The bed can be divided into two regions. In the first the hot air is the dominant mode of heat transfer and the heating is intense. In the second region the dielectric energy alone provides the heat for evaporation and, since the bed temperature is greater than that of the air stream, indirectly heats the airstream which emerges from the bed effectively saturated, $Y_{GOUT} > 0.9\ Y_{SAT}\ (T_{GOUT})$.

As the value of N_T, the number of heat transfer units, increases the width of the first region decreases. The particles at the inlet end of the dryer will dry out first, regardless of any variation of q with moisture content, and a dried zone gradually spreads through the bed.

The fact that the air emerges from the bed effectively saturated allows the overall energy balance equation, assuming quasi-steady conditions (see equation 5.13) to be used directly to calculate T_{GOUT}, the rate of moisture removal and the proportion of input energy used in evaporation. This is done by setting $\phi_G = 1$ in the equation for the relative humidity, equation (5.14). If the outlet gas temperature is taken as being indicative of the solids' temperature as Figure 13 indicates, then the solution of the overall energy balance provides all the drying parameters of interest without the need to solve the transport equations at each point through the bed. This approach is applicable until the edge of the first zone reaches the outlet. For values of $N_T > 10$ this will apply for most of the drying.

The main variable is now Q/GCp_{GIN} which has an analogous role to qb/h_T in that $Q/GCp_{GIN}T_{GIN}$ is the relative input energy from the electromagnetic field and the hot air. The variation of gas outlet temperature is shown in fig. 14 as a function of Q/GCp_{GIN} for two values of T_{GIN}. It can be seen that for $T_{GIN} \leqslant 120°C$ and $Q/GCp_{GIN} < 500°C$ the outlet gas temperature is less than 70°C. As Q/GCp_{GIN} increases above 500°C the outlet temperature rises and becomes progressively more independent of the value of the inlet gas temperature: the electromagnetic energy is the dominant heat source. The energy available for evaporation, see equation (5.15), can be greater or less than Q depending upon the value of T_{GOUT} corresponding to the inlet conditions. While $T_{GOUT} < T_{GIN}$ the rate of moisture removal is less than but of the same order as the sum of the rates for drying with electromagnetic energy alone and hot air alone. For $T_{GOUT} > T_{GIN}$ the rate of moisture removal is less than that for drying with electromagnetic energy alone. As Q/GCp_{GIN} increases, the ratio of the energy utilised in evaporation to the input energy approaches one.

If the variation of the input power Q is known as a function of the moisture held in the bed then the temporal variation can be found. Usually

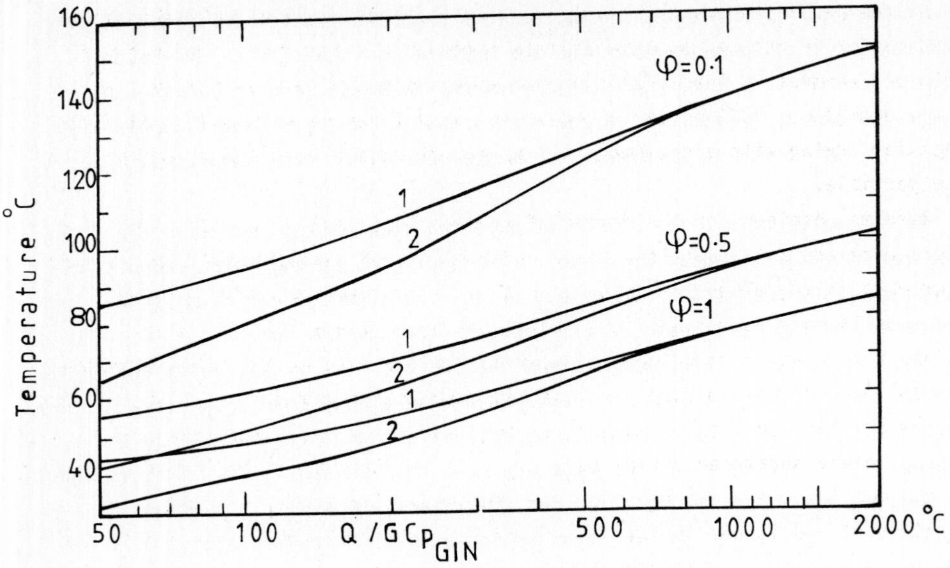

Fig. 14. Calculated values of outlet airstream temperature versus Q/GCp_{GIN}. Curves (1) T_{GIN} = 120°C, curves (2) T_{GIN} = 40°C. Y_{GIN} = 0.005 kg/kg.

this requires measuring the variation of Q. As a first approximation Q can be found from the rate of moisture removal obtained for electromagnetic drying alone as described in the previous section.

(b) Hygroscopic solids with negligible and significant internal resistance to mass flow

As with the previous case, if quasi-steady conditions are assumed the drying characteristics can be obtained from the overall energy balance using the approximation that at the outlet the relative humidity of the airstream is equal to that at the surface of the particles ϕ_{SOUT}. This enables the outlet gas temperature to be found as a function of ϕ_{SOUT} which decreases as the drying proceeds. On this basis the only distinction which can be made between the case of negligible and significant internal resistance to vapour flow is the range of values which ϕ_{SOUT} would take. Temporal variations cannot be found and the modelling must be used in conjunction with small-scale tests.

The values of T_{GOUT} as a function of Q/GCp_{GIN} are included in Figure 14 for two values of ϕ_{SOUT}. For $\phi_{SOUT} \gtrsim 0.5$, $Q/GCp_{GIN} < 500$°C, and $T_{GIN} <$ 120°C, the temperature T_{GOUT} is less than 85°C. As ϕ_{SOUT} decreases below 0.5, T_{GOUT} increases rapidly reaching T_{BP} and above.

As the particles dry out and ϕ_{SOUT} decreases, the energy available for evaporation is reduced as the particle temperatures increase. The rate of moisture removal is less than the corresponding non-hygroscopic case. As ϕ_{SOUT} decreases, so the rate of moisture removal can be reduced to less than that for drying with electromagnetic energy alone due to the heat losses from the particle.

Results obtained for the drying of alumina beads and silica gel with microwaves and a throughflow of air under otherwise similar conditions to the microwave drying alone are shown in Figure 12 (curves marked 2). In each case the quasi-steady assumption is a reasonable approximation as far as the moisture removal is concerned. Comparing the required drying times with the curves for microwave drying alone, it is seen that for alumina the hot air and microwaves both contribute to the evaporative energy resulting in an appropriately shortened drying period. With the silica gel the drying rates in the two cases are roughly the same, the reduction in drying time is not so significant and is due to the absence of a significant heating-up period with the microwave plus hot air drying.

The temperature curves show that the silica gel particles lose heat to the air flowing through the bed and consequently remain at lower temperatures than for the microwave drying alone. The values of temperature obtained, however, are still large. With the alumina, on the other hand, the temperature of the particles remains well below the boiling point temperature T_{BP} for much of the drying, rising to the inlet temperature of the gas towards the end. During these final stages of drying the absorbed microwave power is decreasing and the drying characteristics tend towards those for hot air drying alone.

(c) Influence of particle diameter on fixed bed drying

As the particle diameter decreases so the permissible air flow which can be used without causing fluidisation decreases and at the same time N_T and h_T increase while b decreases.

A typical range of values of N_T and Q/GCp_{GIN} for a 1mm and 0.1mm particle is given by:

1mm $\quad N_T : 5 \rightarrow 10 \quad \frac{Q}{GCp_{GIN}} : 100 \rightarrow 0.5\ °C.$

0.1mm $\quad N_T : (5 \rightarrow 10).10^4 \quad \frac{Q}{GCp_{GIN}} : 10^4 \rightarrow 50\ °C.$

The main consequence of decreasing particle size is that the solids' temperature rises to temperatures approaching T_{BP} as Q/GCp_{GIN} increases, even though qb/h_T is small.

The physical reason behind such changes is that with relatively little airflow the rapid humidification of the airstream to almost saturation values strongly influences the heat and mass transfer from the particles to the airstream, causing the solids' and airstream temperature to rise so that the air is at a high enough temperature at which it can hold the moisture.

(d) Application of the model to a mixed bed of particles

With a mixed bed, the particles dry at the same rate and drying out of one part of the bed before the rest is not seen. In this case the model using the overall energy balance will apply throughout the drying provided the conditions for saturation of the outlet air, that is the value of N_T applies in the first place.

(e) Application of the model to continuous plug flow dryers

The batch situation can be thought of as an elemental portion of the continuous dryer. The passage of this element through the dryer is equivalent to the temporal changes observed for the batch drying.

6. THE PRACTICAL APPLICATION OF RADIO FREQUENCY AND MICROWAVE DRYING

Having considered the basis for r.f. and microwave drying it is now appropriate to discuss the results obtained from the practical application of these techniques in terms of the range of commercial installations, the advantages to be gained, the improved quality of the dried product and the present trend of industrial developments.

6.1 Established r.f. and microwave applications

Examples of commercial applications of r.f. and microwave techniques are shown in table 6.1. In addition to drying, other processes are included for completeness. The improved product quality coupled with the much faster process time has been more than sufficient, in the cases shown, to outweigh the high initial capital cost. Many more applications have been shown to be feasible but were rejected at the time on economic grounds (ref. 23).

TABLE 6.1
Examples of the commercial application of r.f. and microwave heating techniques

APPLICATION	OPERATING FREQUENCY RF	MICROWAVE	TYPICAL INSTALLED POWER kW
Drying			
Sugar	X		30
Flat coils	X		100
Cables made from artificial fibres	X		50
Salt	X		50
Glass fibre "cakes"	X		100
Water based inks and adhesives	X	X	-
Cartons	X		50 - 100
Textile bobbins and packages	X		50 - 100
Textile slivers and large diameter fibres	X		30 - 100
Textile hanks			
Paper	X		900
Vacuum dehydration of fruit powders		X	50
Pasta		X	30 - 60
Tobacco	X		1000
Sand cores	X	X	
Refractory Products	X	X	
Photographic prints	X	X	
Post baking of biscuits	X		100
Womens nylons	X		60
Heating			
Meat rendering		X	30
Precooking of bacon		X	30
Heating wool bales	X		30
Thawing frozen meat	X	X	25
Thawing butter		X	60
Baking of bread	X	X	20
Chipboard manufacture	X		100 - 300
Plastics welding	X		100
Wood glueing	X		25
Preheating of rubber		X	10's
Vulcanisation of rubber		X	10 - 50
Proofing doughnuts		X	25
Curing sand cores	X		60
Wool dyeing	X		100

6.2 Advantages of electromagnetic drying techniques

From an examination of the examples shown above it can be seen that r.f. and microwave drying can offer a number of advantages over conventional methods:-

(a) Improved product quality (see below);

(b) Faster drying times and the possibility of increased production rates e.g. in the textile drying of yarns, drying time of 4-12 hours have been reduced to 15-45 minutes (ref. 24);

(c) More compact equipment, e.g. in pasta drying using a combined hot air/microwave system the overall length of the equipment has been reduced from 120-160 feet to 27 feet (ref. 25);
(d) a cleaner environment, in particular the entrainment of fine particles can be avoided;
(e) the process is capable of automation;
(f) improved sanitation and faster maintenance: the cleaning and overhaul of the smaller sized equipment can be accomplished much faster than for conventional equipment (refs. 25 and 26);
(g) reduction in the wastage as a result of the improved product quality (ref. 26);
(h) reduction in drying costs as a result of improved efficiency and quality;
(i) reliability of the equipment e.g. for paper drying the availability of a new installation during its first year of service was 97% (ref. 26).

6.3 Improvements in product quality

One aspect which cannot be fully appreciated from the modelling is the final quality of the product. Drying with electromagnetic energy has been seen to give the following improvements in product quality:-
(a) the levelling out of moisture profiles (ref. 15);
(b) less migration of additives applied in solution (ref. 21);
(c) in pre-puffing of textured food the expansion of the material due to internal evaporation has prevented subsequent shrinkage and case hardening when the material was dried with hot air (ref. 37);
(d) improved rehydration following drying with r.f. or microwaves owing to the lack of shrinkage (ref. 38);
(e) superior aromatic quality of soluble fruit powders, as compared to spray and conventional freeze drying, due to better retention of the volatile substances (ref. 32);
(f) the production of pharmaceutical granules which are less dusty and more robust than fluid bed dried granules (ref. 36);
(g) in the drying of pasta, infestations of all kinds were reduced, colour was enhanced and the macaroni had a better "bite" (ref. 25).

6.4 Developments in electromagnetic dryer systems

The industrial use of electromagnetic techniques for drying sheets, webs and thick materials in the form of blocks, packages etc., is well established and present efforts are directed more at finding new applications rather than developing new equipment. An exception to this trend is work concerned with combining r.f. and hot air flotation drying of webs into a compact piece of equipment which utilizes the air impingement nozzles as electrodes for the radio frequency applicator (ref. 27). The radio frequency energy, which contributes up to a tenth of the total energy, produces moisture levelling and can, in some cases, promote the internal mass flow within the web to prevent case-hardening thus enabling the hot air drying to operate more effectively than it otherwise would.

The drying of heat sensitive powders and food products is still rather limited. This is due to the fact that conventional methods of drying particulates can be very efficient and because at atmospheric pressure the process temperatures obtained with r.f. and microwave drying are often greater than 50-70°C which for many products is too large. Present developments are aimed at the specialised area of low temperature drying under vacuum to obtain high quality products, and with the combination of fluidised bed and microwave drying as a means of speeding up the air drying.

6.4.1 Fluid bed drying augmented with microwaves

If attrition and gas cleaning are not a problem then the addition of microwave energy to a fluidised bed dryer provides a means of obtaining uniform drying due to the mixing, at increased evaporation rates as a result of the volumetric heating. The equipment can take the form of a multimode oven with the drying compartment of the fluid or spouted bed construction forming part of the cavity, see Figure 15a. Alternatively a travelling wave applicator operating at 896 MHz can be adapted to form a plug flow fluid dryer, Figure 15b. A recent patent (ref. 28) has proposed such a dryer for the final drying of pasta while a spouted bed-microwave combination has been outlined for drying PVC granules (ref. 29).

6.4.2 Freeze drying with microwave energy

The attraction of freeze dried products is their excellent quality - the dried material is very stable and has greatly increased shelf-life. The high capital cost of the equipment restricts the use of this drying to applications where a high quality product is essential.

In freeze drying, frozen materials are placed in a vacuum chamber which is continually pumped to maintain the operating conditions below the triple-point of water ($T = 0°C$, $P_V = 611$ Pa). The product remains frozen and ice is

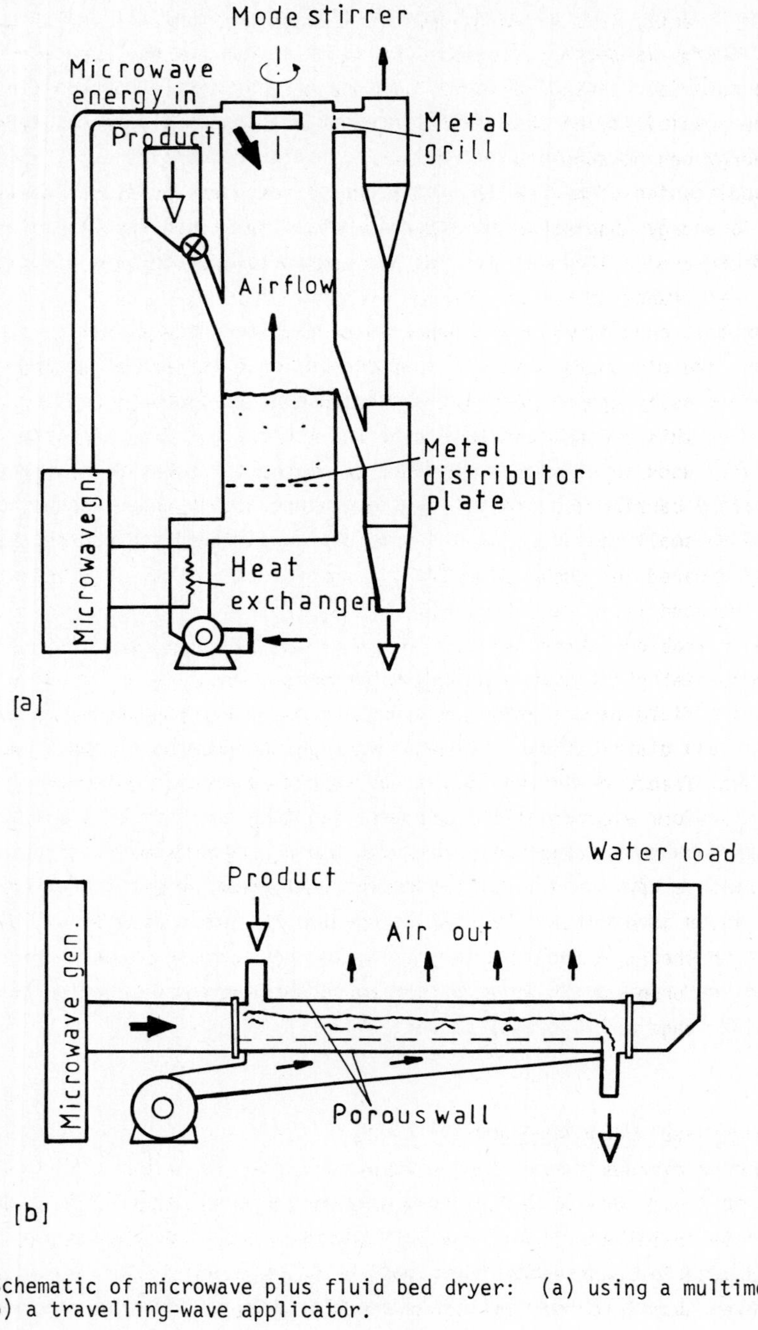

Fig. 15. Schematic of microwave plus fluid bed dryer: (a) using a multimode cavity, (b) a travelling-wave applicator.

converted directly into water vapour. In conventional freeze drying the required energy is supplied by radiant heating. The sublimation front moves into the material leaving the dried surface exposed to the radiation. To avoid the possibility of thermal degradation at the outer surface the rate at which energy can be supplied is limited.

The application of microwaves allows the frozen core to be heated by the volumetric energy absorption into the ice while the dried region is unaffected by the microwaves. The heat and mass transfer process is governed by the internal resistance to vapour flow to the surface of the material. It is important that conditions are maintained below the triple-point since if the ice melts, the microwave energy is preferentially absorbed by the water leading to grossly uneven heating and a breakdown of freeze drying conditions. Modelling of this process requires accurate data if the theory is to be successfully used to predict the operating conditions under which the drying can be safely carried out (ref. 30 and references therein). Experimental results with small scale equipment have shown significantly shorter drying times as compared to radiant drying, for example the drying of sliced pears has been reduced from 18 to 2 hours (ref. 30).

The main problems which have so far prevented the development of equipment up to commercial plant size are ionization of the water vapour which disrupts the electric field distribution and destroys the product, and the uneven electric field distributions obtained in multimode cavities constructed from conventional freeze dryers which did not have the optimum dimensions to produce a uniform electric-field pattern.

One approach to solving these problems (ref. 31) has been to use a travelling-wave applicator situated within the vacuum vessel. The material to be dried is in the form of slabs which are passed down the centre of the waveguide in the region of the peak value of the electric field.

Due to the problems cited above microwave freeze drying is still in the development stage.

6.4.3 Microwave vacuum drying

The use of microwaves for drying under vacuum at higher pressures as compared to freeze drying is, by contrast, much more advanced with industrial size systems installed. This method of drying is used for temperature sensitive materials where the final quality is important. It has been offered as a viable substitute for conventional freeze drying.

The heat and mass transfer characteristics are essentially the same as those for "thick" materials: the moisture is boiled off at the boiling point temperature which is lowered by the vacuum. The range of pressures used is 1-20 kPa (10-200 mbar abs) to give product drying temperatures $< 60°C$.

A number of systems, at different stages of development, are being considered (refs. 32-36). The majority are continuous processes using multimode cavities operating at 2450 MHz to give better homogenity of the electric field and to reduce problems of ionization of the vapour. The microwave input ports must be supplied with vacuum tight windows and the product ports for the continuous systems must be vacuum and microwave tight, Figure 16. At 2450 MHz power is supplied using combinations of 6kW magnetron units, the largest installation available being 96kW with units of 48kW installed commercially (ref. 32).

A different approach using 2 x 25kW magnetrons operating at 896 MHz into four travelling wave applicator waveguides located within a vacuum chamber operating at 10 kPa, has been developed to the pilot plant stage (ref. 34).

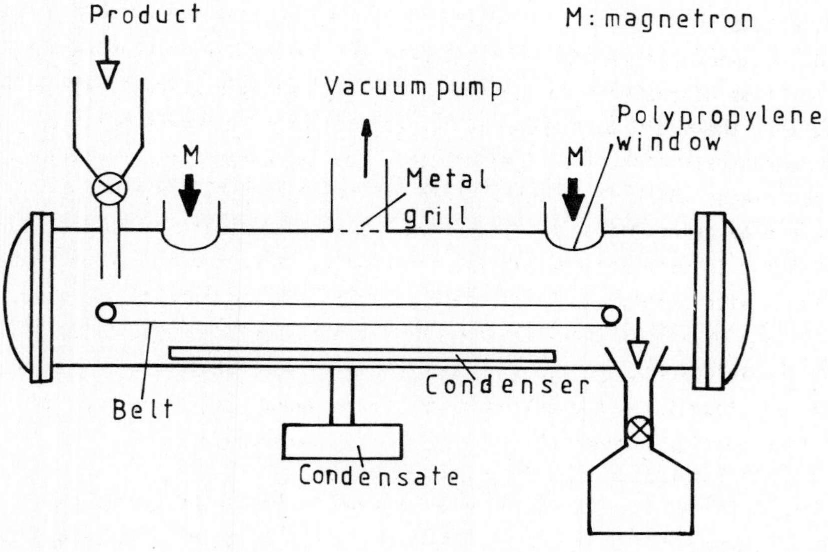

Fig. 16. Schematic of a continuous microwave vacuum dryer.

The water vapour removed from the material is either condensed within the vacuum system and then removed or alternatively at the higher operating pressures it has been possible to use a purge of air sucked through the system by the vacuum pump to take away the vapour to an external condenser (ref. 34). In the case of granule drying (ref. 36) the through-flow of air also prevents caking of the bed.

The material handling systems consist of single or multiple belts to take material in the form of foams, highly viscous slurries, granules or powders. A vertical shute system has been devised for passing cereal grains through a continuous system.

The results obtained from these dryers have been most encouraging with high quality products being obtained in a much faster time with an overall reduction in energy compared to conventional dryers, (ref. 35 for example). Present efforts are directed at scaling up dryers and improving the handling facilities for the materials.

Valuable comments on the development and practical running of such dryers can be found in the papers by Gillet and Lefort (refs. 36 and 34).

7. CONCLUSIONS

Radio frequency and microwave techniques offer the means of overcoming many of the problems of conventional drying, to produce a superior quality product. The fact that there are many industrial applications of these techniques clearly demonstrates that the higher initial capital costs can be outweighed by the advantages which these methods offer. The techniques of electromagnetic energy generation and application are well established and can be adapted to fit in with most processes. The general characteristics of drying can be found with relatively little data so that the applicability of these methods to the drying of a particular product can be readily assessed.

Present developments are aimed at augmenting conventional drying with r.f. and microwave techniques and advancing microwave vacuum drying methods to produce high quality products.

8. ACKNOWLEDGEMENTS

This work is published by permission of the Electricity Council Research Centre.

9. PRINCIPAL SYMBOLS

A	Area	m^2
A_P	Pressure permeability	m^2/s
Bi	Biot number	-
C_a	Mass capacity of gas	Pa^{-1}
Cp	Specific heat	J/kg°C
Cps	Specfic heat of skeleton solid	J/kg°C
D	Overall transport coefficient for internal mass flow	m^2/s
D_L, D_V	Effective "diffusion" coefficient for internal mass flow in the liquid and vapour phases respectively	m^2/s
E	Electric field strength	V/m
E_{max}	Maximum electric field strength	V/m
G	Mass flow of airstream	kg/s
H_B	Binding energy	J/kg
M	Molecular weight	kg/mol
N_T	Number of transfer units	-
P	Total pressure	Pa
P_V, P_{VS}, P_{VG}	Partial pressure of vapour; at the surface of the wet material, in the air stream	Pa
P_o	Atmospheric pressure	Pa
Q	Total absorbed electromagnetic power	W
R	Pore radius	m
R_G	Universal gas constant	J/mol K
T	Temperature	°C
X	Moisture content (dry basis)	kg/kg
X_L, X_V	Mass concentration of liquid (L) or vapour (V) divided by γ_o	kg/kg
V	Volume	m^3
Y	Humidity	kg/kg
Z_o	Dryer length or bed depth	m
a	Surface area per unit length (or bed depth) of dryer	m^2/m
b	Volume to area ratio	m
2d	Thickness of material	m
e_V	Ratio of vapour flow to total moisture flow	-
f	Frequency of applied electromagnetic field	Hz
h_T	External heat transfer coefficient	$W/m^2{\circ}C$
k_M	External mass transfer coefficient	$mol/m^2\ s$
j_M, j_{MS}	Total mass flux; total mass flux at the surface of the wet material	$kg/m^2\ s$

j_L, j_V	Mass flux by liquid and vapour flow respectively	kg/m^2 s
k	Thermal conductivity	W/m°C
q	Electromagnetic power absorbed per unit volume	W/m^3
r	Latent heat of evaporation	J/kg
z	Spatial coordinate	m
γ_o	Density of skeleton solid	kg/m^3
δ_P, δ_T	Flux coefficients	Pa^{-1}, $(°C)^{-1}$
ε	Dielectric constant	-
ε''	Loss factor	-
λ_o	Free space wavelength	m
ρ	Density	kg/m^3
σ	Electrical conductivity	S/m
ϕ	Relative humidity	-
β_T	Correction factor for heat flow	-
Φ	Porosity	-
μ	Viscosity	kg/m s

SUBSCRIPTS

BP	Normal boiling point of the moisture
G	Gas (airstream)
GIN, GOUT	Gas conditions at the inlet and outlet of the dryer
L	Liquid
M	Mass
OUT	Outlet conditions of dryer
P	Pressure
S	Conditions at surface of material (solid)
SOUT	Conditions at surface of material and outlet of dryer
SAT	Saturated conditions
V	Vapour
VSAT	Saturated conditions for vapour

10. REFERENCES

(1)	Dittrich H. F.	Tubes for RF heating, Philips pub. Dept. Einhoven (1971)
(2)		Electrowarme Theorie und Praxis Verlag. W. Girardet, Essen (1974)
(3)	Metaxas A. C. & Meredith R. J.	Industrial Microwave Heating Peter Peregrinus, London (1983)
(4)	Kraszewski A.	J. Micro. Power 12(3) 215-222 (1977)
(5)	von Hippel A.	Dielectric Materials and Applications, MIT Technology Press (1954)
(6)	Kraszewski A.	J. Micro. Power 15(4), 298-310 (1980)
(7)	Lind H. & Popert F.	Brown Boveri Review 51(10/11) 701-711 (1964)
(8)	Akhtarzad S. & Johns P. B.	Electronics Letters 11(24) 599-600 (1975)
(9)	Meredith R. J.	Industrial Applications of Microwave Energy, ed. R.B. Smith, International Microwave Power Institute (Europe) 1974
(10)	Luikov A. V.	Heat and Mass Transfer in Capillary Porous Bodies Pergamon Press, Oxford (1966)
(11)	Bird R. B., Stewart W. E. & Lightfoot E. N.	Transport Phenomena Wiley, New York (1960)
(12)	Eckert E. R. G. & Drake R. M.	Heat and Mass Transfer McGraw Hill, NY (1959)
(13)	Mujumdar A. S. (editor)	Advances in Drying, Vol. 1, Hemisphere Publ. Corp. N.Y. (1980)
(14)	Perkin R. M.	Int. J. Ht Mass Transfer 23 687-695 (1980)
(15)	Jones P. L.	Heat and Mass Transfer in a Radio-frequency Dryer PhD thesis, Loughborough Univ, England (1981)
(16)	Cross A. D., Jones P. L. & Lawton J.	Trans. I. Chem.E. 60, 67-74 (1982)
(17)	Lefeuvre S.	Phys. Technol. 12 155-169 (1981)
(18)	von Czepek E. & Sparckmann H.	Electrowarme International 26(12) 440-447 (1968)

(19) Preston M. D. — 28th Annual Conf. (IEEE) of Electrical Eng.
Problems in the rubber and Plastics Ind. 59-64 Akron Ohio (April 1976)

(20) Lyons D. W., Hatcher J. D. & Sunderland J. E. — Int. J. Ht Mass Transfer 15(5) 897-905 (1972)

(21) Pendergrass J. E., Hatcher J. D. & Lyons D. W. — J. Micro Power 7(3), 207-213 (1972)

(22) Perkin R. M. — Dielectric drying of a fixed bed of particles
ECRC memorandum (to be published) 1982

(23) Hulls P. J. — J. Micro Power 17(1) 29-38) (1982)

(24) "Radio frequency textile drying pays dividends" Electrotechnology 46-47 April 1977)

(25) Smith F. J. — Microwave Energy Applications Newsletter XII(6) 6-12 (1979)

(26) Grassman H. C. — Wochenbl. Papfabr 107(17) 661-665 (1979)

(27) Jones P. L. — UK Patent Appl. No. 8214289 (1982)

(28) Buehler A. G. — "Method for drying pasta products and apparatus for bulk material treatment" UK Patent specification No 156054 published 6 February 1980

(29) ICI Ltd — "Process for drying vinyl chloride polymer wet cake and drier therefore" UK Patent application, GB 2049899 A published 31 December 1980

(30) Arsen H. B. & Ma J. H. — Drying '80 258-264 Hemisphere Publishing Corp. NY (1980)

(31) Tetenbaum S. J. & Weiss J. A. — Proc. Southeastcon Reg 3 Conf. 81 816-819 Huntsville, Ala (April 5-8 1981)

(32) Food Engineering 78-79 February 1979

(33) H. Balfour Ltd — "Vacuum drying apparatus" UK Patent Application GB 2071833 A published 23 September 1981

(34) Lefort J. — "Industrial Processing using Radio frequency and Microwave power" Proceedings of seminar British Nat. Comm. for Electroheat, London (10 December 1981)

(35) Gardner D. R. & Butler J. L. — Paper given at 2nd Int. Drying Symposium Montreal (1980)

(36) Gillet J. E. Proc. 3rd Int. Drying Symposium, 397-406 Birmingham (1982)

(37) Huang H. F. & Yaks R. A. J. Micr Power 15(1), 15-17 (1980)

(38) T. S. Kaisha Ltd "Method of making instantly cookable noodles" British patent specification No. 1587 977 (April 1981)

PROGRESS IN FILTRATION AND SEPARATION 1
Edited by R.J. Wakeman

CONTENTS

PROGRESS IN FILTRATION AND SEPARATION 2

Edited by R.J. Wakeman

CONTENTS